THE SWISS WAY
OF
WELFARE

THE SWISS WAY OF WELFARE

Lessons for the Western World

Ralph Segalman

PRAEGER SPECIAL STUDIES • PRAEGER SCIENTIFIC

New York • Philadelphia • Eastbourne, UK
Toronto • Hong Kong • Tokyo • Sydney

Library of Congress Cataloging-in-Publication Data

Segalman, Ralph.
  The Swiss way of welfare.

  Bibliography: p.
  Includes index.
  1. Public welfare — Switzerland.
  2. Switzerland — Social policy. I. Title.
  HV353.S44 1986   361.6′09494   85-19153
  ISBN 0-03-004593-2 (alk. paper)

Published in 1986 by Praeger Publishers
CBS Educational and Professional Publishing, a Division of CBS Inc.
521 Fifth Avenue, New York, NY 10175 USA

© 1986 by Praeger Publishers

6789 052 987654321

Printed in the United States of America on acid-free paper

INTERNATIONAL OFFICES

Orders from outside the United States should be sent to the appropriate address listed below. Orders from areas not listed below should be placed through CBS International Publishing, 383 Madison Ave., New York, NY 10175 USA

Australia, New Zealand
Holt Saunders, Pty, Ltd., 9 Waltham St., Artarmon, N.S.W. 2064, Sydney, Australia

Canada
Holt, Rinehart & Winston of Canada, 55 Horner Ave., Toronto, Ontario, Canada M8Z 4X6

Europe, the Middle East, & Africa
Holt Saunders, Ltd., 1 St. Anne's Road, Eastbourne, East Sussex, England BN21 3UN

Japan
Holt Saunders, Ltd., Ichibancho Central Building, 22-1 Ichibancho, 3rd Floor, Chiyodaku, Tokyo, Japan

Hong Kong, Southeast Asia
Holt Saunders Asia, Ltd., 10 Fl, Intercontinental Plaza, 94 Granville Road, Tsim Sha Tsui East, Kowloon, Hong Kong

Manuscript submissions should be sent to the Editorial Director, Praeger Publishers, 521 Fifth Avenue, New York, NY 10175 USA

CONTENTS

Page

LIST OF TABLES   vii

LIST OF FIGURES   ix

PREFACE   xi

Chapter

1   AMERICAN AND EUROPEAN WELFARE: UNIN-
     TENDED EFFECTS OF HUMANE IMPULSES   1

     The U.S. Welfare-Dependency Problem   1
     The Contemporary European Experience with
         Poverty   17
     References   38

2   SWITZERLAND AS A WELFARE STATE   48

     References   62

3   SWISS SOCIAL INSURANCE AS A CONSTRAINT ON
     WELFARE DEPENDENCY   66

     The Old-Age and Survivors' Program: The First
         Pillar   71
     The Compulsory Occupational Retirement Scheme:
         The Second Pillar   74
     Invalidity Insurance   76
     The Supplementary Benefits of the Old-Age, Surviv-
         ors', and Invalidity Programs (Ergänzungsleis-
         tungen Provision)   78
     Sickness Insurance and Accident Insurance   79
     Unemployment Compensation   82
     Family Allowances   83
     References   90

Chapter		Page
4	SWISS IMMIGRATION POLICY AS A CONSTRAINT ON WELFARE DEPENDENCY	93
	References	105
5	THE SWISS LOCAL WELFARE SYSTEMS	107
	Zurich	123
	Bern	130
	Geneva	135
	Basel	140
	Aarau	142
	Summary	147
	References	148
6	INDIRECT CONTROLS OF WELFARIZATION IN SWITZERLAND	150
	The Swiss Family	150
	The Swiss Community	160
	Schooling	161
	The Community Youth Authority	164
	Employment	164
	Military Service	165
	Swiss Domestic Relations Court	167
	Psychiatric Facilities	170
	Family Policy	171
	Swiss Welfarization	172
	Counter Factors	175
	References	182
7	CONCLUSION	186
	Lessons for the Western World	186
	Are There Lessons in the Swiss Example?	193
	References	197
INDEX		199
ABOUT THE AUTHOR		207

LIST OF TABLES

Table		Page
1.1	Interactive Welfare Dependency	14
2.1	Income Tax as a Percentage of Gross Earnings across Countries	59
3.1	Comparison of Swiss Social Insurance with Public Assistance, by Item and Function	67
3.2	Annual Full Pension Plan	73
3.3	The Swiss Social-Insurance System: Income, 1980	85
3.4	The Swiss Social-Insurance System: Expenses, 1980	86
3.5	Swiss Special Insurance Systems and Allowances Income: Income, 1980	87
3.6	Swiss Special Insurance Systems and Allowances Income: Expenses, 1980	88
5.1	Swiss Public Assistance Programs: Income, 1980	121
5.2	Swiss Public Assistance Programs: Expenses, 1980	122
6.1	Comparison of Alcohol Consumption across Countries	177

LIST OF FIGURES

Figure   Page

1.1   AFDC Families as a Percentage of All Families
        in the United States   8

1.2   Plot of Percentage of White Persons and White
        Female-Headed Families below the Poverty
        Level   9

1.3   Plot of Percentage of Black Persons and Black
        Female-Headed Families below the Poverty
        Level   10

1.4   Plot of Illegitimate Births per 1,000 Live Births,
        White and Black Women   11

5.1   Social and Individual Help of the Welfare Agency
        of the City of Zurich   128

PREFACE

As a neophyte social worker after the Great American Depression, I was driven by the need to help people. Unlike most social workers, who were satisfied to provide direct assistance, I sought to help people to help themselves. I found that for some, this was possible. Others, aside from some temporary assistance, did not need anyone's help and soon resumed a life of self-sufficiency. But for some, whatever aid was available, public or otherwise, became a habit on which they began to depend. I was challenged to find out why this was so. I was also challenged to determine whether this was true in other lands and with other people. Why is it, I asked, that public welfare becomes a trap from which some poor never escape? Is it the social system that locks them in? Or do they lock themselves in? Or both?

I spent many years studying the problem. As a social psychologist, I examined the human dynamics of welfare dependency. As a researcher, I sought to determine how many remained in the dependency trap. As a researcher, I wondered why so few scholars, public welfare professionals, and social sorkers were willing to examine the issue. In time I learned that this was an area of research avoided by many for fear that an admission of the problem would somehow harm the programs of aid to the poor.

As I have noted in Chapter 1, the problem of transmitted poverty and intergenerational welfare dependency exists to a greater or lesser degree in most of the Western nations. Only in Switzerland, the land that does not claim to be a welfare state, did I find no transmitted welfare dependency. Why this is so and what Switzerland has to offer other Western nations in this regard is presented in the following pages. In my opinion, the other Western nations have put their funds and resources into the provision of income guarantees and supports, while Switzerland has put its energies and resources into prevention of poverty, into the building of self-support incentives, and into preparation for employability and work competence in its population. Where aid is required, it is provided locally, carefully, and with concern for the effect of such aid on the beneficiaries.

In the United States for the past 50 years, legislative and court actions have weakened family and community as well as individual responsibility, while only slightly increasing the rights (without the responsibilities) of the individual and immensely increasing the power and responsibility of the central government. Only now, half a century later, can we discern that whereas central government is sup-

posedly "everybody," everyone's responsibility turns out to be no
one's responsibility. Only now, after having put so many eggs in the
central basket, do we discover that the central basket is excessively
costly, inefficient, frequently incompetent, inhumane, and full of unintended and perverse consequences. Only in Switzerland is the coun-
terforce to centrality present: the principle of subsidiarity in the
national constitution and in daily practice at all levels.

    The conclusions arrived at in this work have recently been con-
firmed by yet another Swiss scholar. Walter Rüegg, in his chapter
on Switzerland titled "Social Rights or Responsibilities?" in the
Eisenstadt and Ahimer report, <u>The Welfare State and Its Aftermath</u>,
indicates that Swiss success in social insurance and welfare effectiveness derives from a high degree of decentralization, which makes
each community and neighborhood responsible for itself (pp. 188-89).
Rüegg supports my conclusions that the elderly and handicapped in
Switzerland are generally well provided for: "Their average wealth
amounts to more than double that of those still working" (p. 196).
He presents a view of Swiss schooling that is unlike that of other
lands in that education is seen as a laboratory of civic and social responsibility rather than a mere service to children and their parents
(p. 185). Rüegg describes the Swiss tradition of the state "being
fundamentally the direct responsibility of free citizens in the political,
public, and social order" (p. 183). This tradition of citizen participation in public affairs follows naturally on the citizen's duty to mil-
itarily protect his community, his canton, and his nation. Rüegg
views the phenomenon of extensive mutual, neighborly aid and self-
help in the Swiss community as a realization of a society in which
individuals are both secure and yet free to pursue the activities chosen by themselves (pp. 196-97). This, he believes, is a product of
social responsibility rooted in primary institutions, with only subsidiary functions relegated to secondary institutions. In short, the
Swiss have learned that only the local community can effectively and
fairly manage community welfare and social control in a manner that
is responsible and that promotes social responsibility and individual
autonomy. The Swiss have accepted the principle of social solidarity
that citizens bear responsibility for one another; but they reserve the
right not only to help their neighbors but also to participate in helping
them avoid dependency in the future.

    "Classical liberal thought," according to a <u>New Republic</u> edi-
torial (1985, p. 8), "puts a primacy on personal liberty up to a point
where an individual's actions threaten to harm society as a whole."
The combination of state welfarism and the effects of a subsidized
program of single-parent families with inadequate socialization of children has produced a society of high levels of violent crime, job-
unready youth, high drug and alcoholism rates, and related ills. It

is apparent that in the Western world it is not the individual actions of citizens that are at the core of social problems. On the contrary, it is the inaction of citizenry in their unwillingness to retain and maintain the community responsibility necessary for a functioning democracy. Without community participation and responsibility, public order and public welfare can only be maintained on a centralized basis, with a heavy police and social-control contingent and a citizenry at the mercy of an impersonal and generally unaccountable centrality.

I want to express my thanks to the many scholars, public officials, social workers, and others in Switzerland and Europe who have shared their information with me.

Scholars:

Dr. Walter Ackermann, Hochschule-Sankt Gallen
Dr. Roger Girod, University of Geneva
Dr. Ruth Gurny, University of Zurich
Dr. Erich Gruner, University of Bern
Dr. Jürg Hauser, University of Zurich
Dr. Mattias Hauser, University of Zurich
Dr. Hans-Joachim Hoffmann-Novotny, University of Zurich
Dr. J. Kellerhaus, University of Geneva
Dr. Martin Killias, University of Lausane
Dr. Kurt Lüscher, University of Constanz
Dr. Nicola Madge, London School of Economics and Political Sciences
Dr. Charles Ricq, University of Geneva
Dr. Gaston Schaber, Pedagogic Institute, Luxembourg
Dr. Willy Schweizer, University of Bern
Dr. Iwan Seewandono, University of Rotterdam
Dr. Sylva Staub, School of Social Work, Zurich
Dr. Heinz Strang, University of Heldesheim
Dr. Hans Tuggener, University of Zurich
Dr. Henri Zwicky, University of Zurich

Public Welfare Officials:

Mr. Jean Marc Boillat, Social Service, City of Geneva
Dr. Michael Horn, Fürsorgeamt, City of Bern
Mr. Guy Perrot, Hospice Général, Canton of Geneva
Mr. Peter Tschümperlin and staff, Socialamt, City of Aarau
Mr. Victor Stohler, Allgemeine Sozialhilfe, City of Basel
Dr. Paul Urner, Fürsorgeamt, City of Zurich
Ms. Regula Wagner, Fürsorgeamt, City of Zurich

Youth Authority Officials:

Mr. Peter Hug and social workers of the Jugendamt, District Horgen,
    Zurich Canton
Dr. Victor Reidi, Jugendamt, City of Bern
Dr. Emil Weinmann, Jugendantwaltschaft, District Horgen, Zurich
    Canton

Other Public Officials:

Dr. Til Bandi, Bundesamt für Sozialversicherung, Bern
Judge Alexander Brunner, District Court, Zurich
Mr. Jean-François Charles, Bundesamt für Sozialversicherung, Bern
Mr. Karl Graf, Gemeinderat, City of Zurich
Mr. Urs Hadorn, Bundesamt für Polizeiwesen, Bern
Dr. Hans Peter Tschudi (former Bundesrat), Basel
Dr. W. Zingg, Bundesamt für Statistik, Bern

Other Agencies:

Mr. Edvard Batliner, Regionaler Jugendberatungsdienst, Aarau
Mr. Eugene Brand and Mr. Oliver Wuillemin, ATD Vierte Welt,
    Tréyveau
Dr. Joseph von Duss, Institute für Ehe und Familie, Zurich
Mr. Karl Eisenring and Mr. Vigeli Venzin, Regionales "Drop In"
    Drug Counseling Center, Aarau
Mr. Jurg Frauenfelder, Informationsstelle für Sozialwesen, Zurich
    Canton
Mr. Heinz Glattfelder, Pro-Familia, Lucerne
Mr. Caspar Nachman, Pro-Juventute, Zurich

Others:

Mr. and Mrs. Manfred Brennwald, Udligneswil
Mr. and Mrs. Dieter Hoffmann, Schlieren
Dr. Jürg Meyer, Basler Zeitung
Dr. Reudi Rentsch, Olten
Mr. and Mrs. Hyman Yantian, Glattbrugg

    In addition, I wish to express my appreciation to Mr. Hyman
Yantian of Glattbrugg for his gracious help in translation of selected
documents and to Professor Willy Schweizer, University of Bern, for
his careful reading of the manuscript and for his generous help and
advice.

I also wish to thank my colleagues, Professors Alfred Himelson, Albert Pierce, and Ernst Weber of the California State University-Northridge, as well as Professor Emeritus Marshall Clinard of the University of Wisconsin-Madison, for their advice and help.

I am also indebted to the small grants division of the California State University-Northridge Foundation for subsidy of some of the costs of translating documents.

If any errors become apparent in the book, they are entirely my responsibility and are not in any way the fault of the many people who served as advisers and sources of information.

REFERENCES

New Republic. 1985. "The Dope Dilemma." New Republic 3, no. 665 (April 15): 7-8.

Rüegg, Walter. 1985. "Social Rights or Responsibilities? The Case of Switzerland." In The Welfare State and Its Aftermath, edited by S. N. Eisenstadt and Ora Ahimer, pp. 182-99. Totowa, N.J.: Barnes & Noble.

1

AMERICAN AND
EUROPEAN WELFARE:
UNINTENDED EFFECTS
OF HUMANE IMPULSES

THE U.S. WELFARE-DEPENDENCY PROBLEM

Welfare in the United States is a multifaceted problem. It is an economic problem in that it represents an expanding dollar cost without making a direct contribution either toward the gross national product or toward the rehabilitation of many of the affected population. As such, it is not an insubstantial contributor to inflation. It is a social problem in that it provides an expanding threat to the family as an institution by substituting for the marginal worker and family provider a periodic government welfare check. In the process, it adds impetus and reinforcement to the growth of the female-headed family as the focus of a distorted socialization process for children, with by-products of inadequate valuation of education, work, and career training. These families are primarily located in geographical areas with high juvenile delinquency rates, violent gangs (which serve as substitute child rearers), child abuse and neglect, high crime rates, drug addiction, and other social ills (see Auletta 1982; Banes and Ellwood 1983; Banfield 1968; Murray 1984; Segalman and Basu 1981; Sheehan 1976). The problem is further aggravated in that this population contains a growing component of unmarried mothers and illegitimate children who are themselves products of prior welfare-dependent families (see Berger 1984; Cherlin 1977; Garland 1967; Hill and Ponza 1983; Hulbert 1984; Moore et al. 1979; Morgan 1978; Mott 1983; Nathan 1978; Office of Management and Budget 1984; Rein and Rainwater 1977; Shkuda 1976; Sklar and Berkov 1974; U.S., Department of Health and Human Services 1983). Welfare dependency is also a social problem in that it undergirds the development of a growing cultural gap between the mainstream population and chronically dependent, multiproblem families. It is a political problem in that neither its definition nor its causation can be agreed upon—all because the underlying political issues prevent a rational analysis and the design of a logical solution (see Banfield 1979; Chapman 1977; Kristol 1971). Ellwood and Summers (1985) conclude that the poverty issue is grid locked in the United States (p. 3).

In the past, in Europe and the United States, welfare was seen as a temporary aid to unfortunates who were in need because of circumstances beyond their control (Axinn and Levine 1975; Coll 1969). If a person had to continue to be welfare dependent, society and its agents, welfare workers and their psychiatric consultants as well as teachers, tended to view such persons as lazy, confused, ignorant, incompetent, or individually deviant in some other way (Segalman 1976). This was the structural-functional (consensus) view, which saw society as an interacting system of systems into which people and social subsystems fitted themselves in the creation and maintenance of societal functionality (Parsons 1951). Those who were temporarily in need or those who were in need in great numbers were seen as the victims of dislocations within the social structure, a problem to be resolved by such social-engineering mechanisms as employment development and job retraining.

As the U.S. depression of the 1930s receded, and as unemployment reached negative levels during World War II, observers of the welfare system discovered a steadily increasing welfare-dependent population aside from the aged, the handicapped, and the infirm. This group was estimated to represent 5 percent of the total welfare population in 1950, 10 percent in 1960, and 20 percent in 1970 (Rein and Rainwater 1977), and there are indications that it has reached higher levels in 1980 (see Hill and Ponza 1983; Morgan 1978; Mott 1983). The pattern of a growing cycle of welfare-dependent populations has also been noted by Aldous and Hill (1969), Salamon (1978), Banfield (1969), and Moynihan 1968). Wilson (1980) distinguishes between the lower class, who are the working poor, and the underclass, who depend upon welfare and other marginal resources for subsistence. The costs of the chronic welfare-dependent families, in both financial and human costs, are highly disproportional to the size of the affected population, according to Banfield (1968).

Whether the growth of this population derives from structural, individual, or interactive causes is a matter of debate, but the external press on the low-income family to remain on welfare is considerable. Chilman (1979) lists some of the major pressures:

1. Lack of suitable employment opportunities for unskilled women that also permit child-care arrangements;

2. Aid to Families with Dependent Children (AFDC) policies that make it difficult for natural families to be on welfare;

3. Welfare benefits that exceed potential wages, thus discouraging a provider from remaining as the family support;

4. An availability and level of welfare that influences mothers not to marry a poor prospect, not to stay in marriage, and not to have an abortion; this point is also supported by Butler and Heckman (1976) and Wilson (1980).

5. An AFDC-U program (permitting unemployed fathers to remain at home) that is not universal among the states, and which is often a difficult enrollment;

6. Supreme Court rulings of the 1960s that made it acceptable for a noncontributing "boyfriend," but not an employable husband, of the mother to remain in the home;

7. A Supreme Court ruling in 1970 that made men who assume the role of spouse in AFDC families to be considered not responsible for contributions to the children; thus, it is in the interest of unmarried mothers and fathers either not to marry or to marry someone else.

Entry into the labor force is made more difficult, according to Beeghley (1978), in that the only jobs available to the poor are those that do not pay enough or last long enough to alleviate their poverty.

Banfield (1969) and Gilder (1973, 1978, 1981a) lay great emphasis on the situation of unemployed and marginally employed men who are displaced or relieved by welfare as the mainstay of their own children and wives. The manner in which welfare is administered also takes its toll on the family stability of the poor, according to Stack (1974).

(Mass service) caretaker agencies, such as public welfare are insensitive to individual attempts for social mobility. A woman may be cut off the welfare roles when a husband returns from prison, the army or if she gets married. Thus the society's welfare system collaborates in weakening the position of the (welfare family) . . . male. [P. 113]

In addition to the above, pressures on the poor derive from evolving expectations about welfare to the effect that AFDC is now a right rather than a privilege, that the society is required to assume responsibility for their poverty rather than that poverty is a personal responsibility and concern, and that new definitions of welfare as "income transfer" rather than temporary subsistence aid must provide a "decent" standard of living on a continuing and secure basis (London 1976). London quotes Homans on the psychological effect of the definition of aid having changed from "privilege" to "practice" to "right" to "entitlement." "The more often under a particular stimulus an activity is rewarded, the more anger (and psychological disturbance) will be displayed when the same activity (in this case—welfare dependency) is emitted without its reward" (Homans 1950, chap. 17). Thus, the situation of the welfare-dependent poor is a social and psychological structural bind constructed by welfare policy, welfare administration, the employment situation, court rulings, and the poor's own understandings of welfare function. Gilder, in the "Firing Line" broadcast of February 8, 1981 (1981b, p. 9), describes this combination of forces as "almost a matter of entrapment" in the form of benefits leading many of the poor into increased family breakdowns, youth unemployment, and crime. With the combination of distorted and deviant constraints and options facing the long-time poor, it becomes quite understandable why so many of these families elect to continue on AFDC support.

Rutter and Madge (1976), in their review of research on intergenerational poverty, refer to the so-called poverty trap, in which low income, welfare operations, and a wide-ranging multiplicity of problems tend to keep some families from moving out of welfare dependency. Sheehan (1976), in her New York City study, A Welfare Mother, reported that the family exhibited a number of characteristics that have been associated with residual welfare dependency. These characteristics include

1. A matriarchal pattern of family structure;
2. Many children;
3. Early motherhood;
4. Only a tenuous relationship between the family and (a) community, (b) religious organizations, (c) community resources, (d) schools, and (e) centers of community services and a general fear and suspicion of agencies of community responsibility;
5. Only a tenuous relationship between the immediate family and relatives in the extended family;
6. A maternal outlook that expresses no hope of or aspirations for eventually getting her children out of poverty either by seriously supporting efforts toward their education or by encouraging their acquisition of vocational skills;
7. A maternal child-raising orientation that lacks authority and purpose and that is generally disorganized and confused (that is, mother's control of the children is usually ineffective and dependent on episodic physical punishment, which fails to work when the children become physically larger than the mother);
8. A fatalistic point of view;
9. A present-time orientation that promotes impulsive behavior and short-attention-span activity, except in the matter of television soap opera programs;

10. A history of poor marriages or liaisons;

11. Behavior patterns in older children that indicate a continuation of welfare–dependent patterns (for example, one daughter was already a welfare recipient with her own child, and a son already had an expensive heroin habit);

12. A history of missing or unemployed husbands or fathers of the children, each of whom had had only a fluctuating, temporary place in the family constellation;

13. A disorganized and impulsively operated home, with little meal planning and very little organization of familial duties;

14. A pattern of financial management and credit misuse that is chaotic and not operated on a basis of purposeful or survival priorities;

15. A relationship with the welfare authorities that is based less on the true facts of the family's situation and more on an attempt to gain maximum financial aid with as little follow–up investigation as possible.

These maternal characteristics serve as a role model for the children that is not directed toward their escape from welfare dependency. Although the behaviors may in many instances be described as attempts to cope with the day–to–day exigencies of life on welfare, these patterns serve to make the children less competent to perform mainstream functions. The children have no aspirations to get out of poverty. The life pattern that they have learned is a model of continuing dependency. Education is devalued by the children and, at best, they are apathetic about schooling. The mother does not hesitate to keep a child out of school to help with child care. Although she may verbally assent to cooperate with the school, her behavior demonstrates to her children a realistic unwillingness to follow the school's suggestions. Her fatalistic attitudes and her belief that she has no control over her situation is a learned pattern and is exhibited in the children's behavior. The lack of a male role model and the lack of an employed role model in the family deprive the children of an opportunity to learn about reciprocity, which is the basis of most relationships in the world of employment and mainstream social organization. The mother's view of herself as unable to change anything in her life provides her with a sense of self–devaluation and negative identity that pushes the older children into the streets in their search for attractive role models.

The mother's behavior tends to keep her and the children welfare dependent. The way the family is handled by the systemic institutions also tends to perpetuate the family's welfare dependency. Instances that exemplify this pattern include activities such as the provision of cash for moving expenses to the client (which was not receipted and, in the process, was used for other purposes); provision of extra food stamps (when the client reported the first supply lost without checking and, in the process, received a double allotment for the period); and acceptance of the client's feeble excuses for non-involvement in the work training program. Thus, the welfare system provides no inducements to the client to change her pattern of dependency on the welfare system, and frequently even rewards her for lying and manipulating the welfare system. The so-called stigma of welfare is obviously not effective in providing incentives to the client to become self-sufficient. If anything, what stigma there is seems to encourage continued welfare dependency and keeps the client from seeking her way into the mainstream community. The occasional sympathy for her having to be dependent on welfare, as expressed by the various eligibility workers, merely reinforces her acceptance of welfare as a way of life. The agenda for her welfare-system contacts has only one purpose—how to keep up her eligibility status, rather than how to establish her family as members of the productive and participative marketplace. In the midst of the depression of the 1930s, Ruth Smalley, in a Family Service Society of America professional pamphlet, emphasized the fact that welfare interviews that did not focus on long-run client rehabilitation merely reinforced the client's remaining as a residual welfare-dependent family. The pattern of eligibility contacts without concern for movement away from dependency has now apparently become universal in most welfare interviews.

The plethora of workers known to the client, all of whom have only short and occasional contacts with her, prevent the client from really getting to know and trust any one worker. This situation also prevents any one worker from really getting to know the client and her family. Because of the pervasive turnover in workers, many of whom do not really make a career of working within the welfare system, the name of the game for each worker is not to seek to work with the client on her long-range problems but, rather, to quickly resolve whatever has surfaced for the moment. Because this type of client has difficulty in understanding relationships with an institution and can change at this point only on a basis of personalism, the lack of contact with the same worker over the years indicates a particular failure on the part of the welfare system to have any meaningful, helping impact on the client to "get her act together," whether for the purpose of becoming self-sufficient, or even for the more limited purposes of better ordering her affairs and gaining more control over her life activities.

Contacts with the schools also do little to move the children toward eventual self-sufficiency. A self-fulfilling prophecy occurs in that the client in A Welfare Mother does little or nothing to prepare her children for a learning experience in the schools, and the schools match this with an expectation that the children will fail. Thus, the children can be said to be both "dropouts" and "push-outs."

Under any analysis, A Welfare Mother is a clear demonstration that there are interactive, almost symbiotic forces that serve to make and keep the client dependent on the system as a providing and generally permissive parent. This view of the welfare dependent has been supported by Ford (1962), Pruitt and Van de Castle (1962), Anderson (1965), and Stone and Schlamp (1965).

The report by Sharff (1981) of an anthropological study of welfare families in East Harlem also provides support for an understanding of how the coping mechanisms of residual poverty, while quite understandable from the viewpoint of the particular family, can become dysfunctional for any move toward reunion of the family with the mainstream of the society, even when viewed from a transgenerational orientation. In the East Harlem study, it was found that the female-headed, matriarchal, welfare-dependent family, seemingly without planning to do so, directs its children toward specific coping models. Examples of such direction include

1. An elder son who is encouraged to free himself from school requirements as early as he can so that he may become the street representative of the family: He helps to protect the family from harm by his gang alliances and helps to supplement the family's disposable income by contributing his street earnings and his share of gang booty.

2. A home and child-care manager—a daughter who takes over the care of the younger children and who, in turn, and in time, becomes pregnant and brings in her share of welfare support: As in the case of the older son, this child is also encouraged to free herself from school ties as soon as possible so that she may free her mother of home duties as soon as she can.

3. A family advocate, usually a son but sometimes a daughter, who is encouraged to stay in school: This is the child who accompanies the mother in all contacts with authorities, whether welfare, school, or police. This child learns to present the family's needs to authorities, to ask for services and benefits, and to protect the family in case of trouble with authorities.

4. The development of understudies for each of these roles, as younger children become available: This is particularly important in the case of the street representative, who cannot be expected to be available indefinitely in view of the hazards of gang conflict, delinquency, and periodic incarceration. Similarly, as other younger girls become available, they learn the roles of welfare motherhood.

The tie between welfare dependency and hard-core delinquency has been reported by Walter Miller (1958). Psychological dependency in welfare clients was reported by Kluckhorn (1958) and Schneiderman (1974). Battle and Rattner (1963) and Herzog (1963) found values and norms among the welfare-dependent similar to those found by Sheehan (1976).

The work of Sanford M. Dornbusch et al. (1985) indicates that mother-only families, when compared with households with two natural parents, are significantly less effective in control and socialization of adolescent children, and particularly of male children. Whether this is a lack of surveillance, lack of appropriate teaching, or lack of social support for the single parent is not yet known, but the effect is seen in the higher level of deviance among such children. It is clear that this is a particular concern for single mothers who are adults and who have had adequate social and employment experience and who are on temporary assistance. The effect of this weakness in single-parent, welfare-dependent families where the mother is very young or has herself never been adequately socialized for self-sufficiency is probably compounded.

FIGURE 1.1

AFDC Families as a Percentage of All Families in the United States

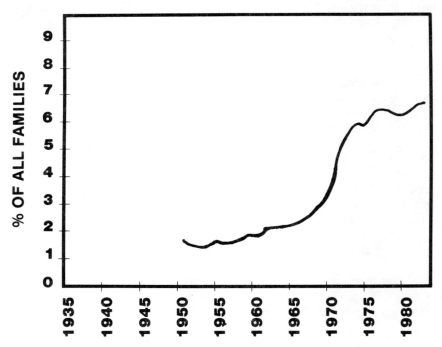

Source: Figure produced by Professor Albert Pierce, Department of Sociology, California State University, 1985.

FIGURE 1.2

Plot of Percentage of White Persons and White Female-Headed
Families below the Poverty Level

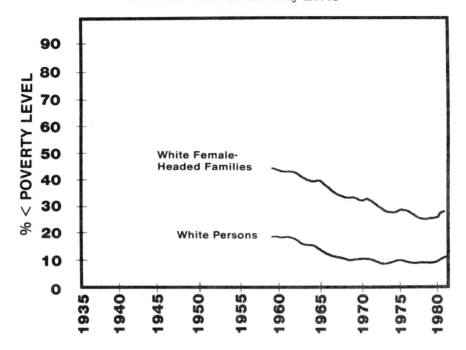

Source: Figure produced by Professor Albert Pierce, Department of Sociology, California State University, 1985.

     Evidence of the growth of AFDC from 1950 to 1980 is presented
in Figure 1.1. This growth in welfare occurred at a time during
which poverty declined in white families generally and even in white
female-headed families (see Figure 1.2). Poverty also declined in
black families generally and in black female-headed families (see
Figure 1.3). If both black and white family poverty and even female-
headed family poverty declined during this period, who made up the
growth of the AFDC welfare population? We find a clue for this in
Figure 1.4, which demonstrates the continuing growth of child illegitimacy rates during the same years. Thus, it would not be a weak
hypothesis to suggest that there is welfare in the United States that is
either caused or abetted by patterns of illegitimacy, by child nonsupport by fathers, and the availability of AFDC under its current

FIGURE 1.3

Plot of Percentage of Black Persons and Black Female-Headed
Families below the Poverty Level

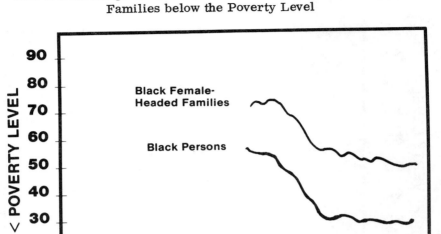

Source: Figure produced by Professor Albert Pierce, Department of Sociology, California State University, 1985.

regulations and delivery patterns. (For comparison of U.S. and
Swiss illegitimacy rates, see the beginning of Chapter 6.)

Not by any means can the residual poor be viewed as the only
model of poverty on the U.S. scene. There are three levels or models
of poverty that should be considered. These are the transitional poor,
the marginal poor, and the residual poor.

The transitional poor are a population whose experience with
poverty is only temporary, covering a relatively short period of
time. In most cases, this population climbs out of poverty in a short
while. This condition is often precipitated by a brief spell of unemployment with limited or no compensation, expensive medical problems, expensive legal litigations, or individual incapacities that make
employment impossible. This group consists of immigrants, both
those of the early 1900s and the contemporary Southeast Asian, Cuban,

and other refugees who, because of cultural, language, or other as-similative adjustments, are currently provided with temporary aid (although aid to refugees by the government is only a recent development for immigrants).

It should be remembered that most of these populations sooner or later are able to become adequately self-sufficient themselves or are able to provide their children with a basic foundation for socio-economic upward movement.

At various time periods, the marginal poor may include the transitional poor. This population also contains many of the working poor who earn just enough to provide for their basic subsistence needs or, sometimes, a bit better. From the daily subsistence point of view, this population is economically marginal. A downturn in the national employment picture or a family mishap can tilt their situa-

FIGURE 1.4

Plot of Illegitimate Births per 1,000 Live Births,
White and Black Women

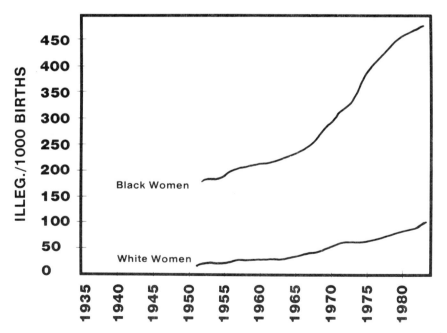

Source: Figure produced by Professor Albert Pierce, Department of Sociology, California State University, 1985.

tion toward a condition of dependency. Some climb out, while a proportion of the population may succumb to residual poverty. It is with
such families that a continued period of employment, a rising level
of expectations, and a willingness to invest themselves in their children's education will help them into secure self-sufficiency.

Residual poverty is defined as the condition of a person or family that remains in poverty through a continued period of time. Most
often, the poverty is transgenerational for this population. Their
children and, in some cases, their grandchildren are subsidized by
welfare grants. This residual poverty population, unlike other welfare clients, maintains a life pattern that is generally unrelated to
mainstream employment. Instead, it interacts symbiotically with the
welfare system and the urban ghetto culture. This group becomes
locked into continued dependency through a combination of factors related to its own life patterns, the social attitudes of welfare stigma,
and the operation of welfare and related institutions. According to
Rein and Rainwater (1977), this group represents more than 60 percent of welfare expenditures, although it is still only a minority of
the welfare population.

During the period when the residual welfare population began
to geometrically expand, a new view about welfare became ascendant
in the United States. This view, of course, had little relationship to
the facts about welfare, but it gained popularity on the coattails of
the U.S. youth rebellion of the 1960s. This approach to welfare no
longer viewed it as a temporary aid, which assumed that self-sufficiency was the reciprocal expectation for all individuals in the society.
Instead, it viewed welfare as a mechanism for income redistribution,
by which the rich would become more equal to the poor (see Bedeau
1967; Durham 1973; Feagin 1975; Piven and Cloward 1971; Ryan
1971). This view interpreted need in a simplistic manner, regardless
of whether or not the individual could have avoided the problem that
created the need in the present, regardless of whether the individual
needed to be helped to prevent becoming dependent again, and regardless of whether or not children were being taught in ways that would
make them also welfare dependent in their adulthood. All this was
considered irrelevant. Welfare was now considered a right and no
longer a temporary, conditional aid. No longer was the client required to seek rehabilitation and self-sufficiency in order to end dependency. The client was now seen as the victim of a societal conflict in which the rich and powerful organize the society and keep the
poor in their place. Nothing was expected from the client; everything
was expected from the society (Dahrendorf 1958).

Thus, in the United States welfare became a divisive issue. The
conservatives (and a growing number of the majority who make up the
middle class and who carry most of the load) take the position of con-

demning all welfare other than for the aged and the handicapped. These are opposed by the liberals, joined by the social work profession, the minority and ethnic movements, the welfare administration bureaucracy, some residual social activists for whom the 1960s rebellion has never ended, and the Marxists, who seemingly hope to bankrupt the capitalist system by the expansion of welfare. An evident supportive group for welfare expansion can be found among women's leadership, many of whom see the welfare programs as a mechanism by which women can more easily escape from or avoid dependency on men, while experiencing the benefits of family life (Wishnov 1973).

Thus, the welfare problem in the United States, with its huge economic costs and even greater social costs in lost life opportunities for the true victims, the children, is seemingly unresolvable as long as it is viewed from one or another monolithic standpoint, and as long as it has its ideological champions from both wings (See Anderson 1978a; Doolittle, Levy, and Wiseman 1977; U.S., Department of Labor, Bureau of the Census 1974). Apparently those who know the welfare clientele least in the United States seem to be among the first to do battle over the subject. Welfare consideration in the United States is a racial issue, a sexist issue, a vested-career issue, a bureaucratic power issue, a pro- or anti-Marxist issue, a legalistic issue, a right-to-teenage-motherhood issue, and many other types of issues; but it is not, unfortunately, viewed in its true light—as an issue of children at risk. The Ford Foundation (1980) describes this population group as

> the underclass—a group of seriously distressed people many of whom are (minorities) (and) whose condition seems to be impervious to traditional attempts at improvement. (This population) is concentrated in central cities . . . suffers much personal misery and accounts disproportionately for strains on family, neighborhood and society, not least through illegitimacy, delinquency and crime. Moreover their problems often weigh heavily in welfare, employment and housing, reducing their effectiveness and impairing public confidence (in these programs) to the detriment of much larger numbers of other disadvantaged people. [P. 5]

A compilation of external and internal factors relating to welfare dependency in the AFDC family is presented in Table 1.1.

The poverty-welfare dilemma is compounded by other social problems in the United States. The low quality of welfare children's socialization at home and education at school is aggravated by interaction with the unintended dysfunctional effects of the welfare system. The failure of many welfare children to achieve academic and voca-

TABLE 1.1

Interactive Welfare Dependency

"Externals," Effects of Chronic Welfare Dependency	"Internals," Effects of Chronic Welfare Dependency	"Behaviors," Effects of Chronic Dependency	Transmitted Value System for Children in Chronic Dependency
(a) Lack of access to quality education for self and children	(a) Sense of powerlessness over one's fate	(a) Impulsivity in purchasing (unearned money is more easily spent)	(a) Children's perception of parents' message that there are no opportunities for them in mainstream
(b) Shrinking expectation of employment opportunities for upper-level labor market	(b) Sense of shrinking roles and skills usable in mainstream contacts and especially in employment	(b) Paying more for less goods	(b) Children's perception of message from parents that welfare as a way of life is acceptable
(c) Shrinking expectations of change and of hope for change	(c) Sense of inability to control one's present or to plan for future	(c) Nonuse of birth control (Who cares?)	(c) Weakening of communication skills for contact with mainstream
(d) Shrinking aspirations for self and children	(d) Lessened interest in, and involvement in, children's learning along with less-	(d) Dropping of behaviors necessary for mainstream involvement	(d) Apartheid social distance from the rest of society becoming the norm

14

(e) Loss of positive validations by others in mainstream

(f) Acceptance of positive validations by others in the dependent subculture of dependency

(g) Dropping of self-reliance and employment role patterns

ened concern for children's attendance and progress in school

(e) Lessened sense of purpose for former family provider

(f) Lessened pattern of respect for purpose of former family provider by spouse

(e) Lessened control of older children who become relegated to peers for regulation and socialization to antisocial life patterns

(f) Early childbearing and delinquency for children

(g) A multiproblem complex of behaviors related to growing family disorganization

(e) Closer ties to, and dependence upon, the subculture of welfare dependence, matched with increased routinization, impersonality, and noninvolvement of welfare personnel in the life of the client family

15

tional skills, upon which their future employability depends, is in great part also compounded by the unintended dysfunctional effects of the welfare system. The lack of well-paying, attractive jobs for unskilled potential workers in welfare families is aggravated by the presence of millions of illegal and legal immigrants who are eager to accept the employment rejected by welfare dependents. Similarly, the current need for at least two wage earners in each family, if the family is to rise above poverty levels, is unfulfilled in the welfare population where single-parent, female-headed families predominate. Finally, the heavy tax load on the working-poor family serves to aggravate the pressures on lower-class families and adds to their difficulties in rejecting family breakup and welfare as a way of life.

This dismal report on U.S. public welfare is not universal. In fact, only a portion of AFDC beneficiaries become residually welfare dependent (see Ellwood and Summers, 1984, p. 32). Most AFDC dependents utilize the program as temporary support, while they rebuild their economic self-sufficiency or else use it as a supplementary aid because of special conditions (many children, handicapped or ill wage earner, and the like). The aged and handicapped beneficiaries of Supplementary Security Income (SSI) are not expected to become self-sufficient but many do. In the case of AFDC, however, chronic welfare dependency is becoming clearly identified. Welfare as a way of life is fast expanding within the recipient population and other populations related to it (boyfriends, grandmothers remaining on welfare as substitute parents, and the like). Ellwood and Summers believe that widening welfare benefits without careful analysis of these populations can have serious disruptive effects. They support the view that assistance should be disbursed by selective categories rather than on a broad income-distribution basis.

Whether the welfare system has caused some families to lose their initiative or whether the loss of such initiative has been triggered by some other factor is less relevant than the fact that welfare has served to delay the return of sizable populations to self-sufficient participation within U.S. society. Whether the welfare system has promoted the breakup of many U.S. families or merely helped these families to remain dependent while the husbands have gone on to create new potentially dependent families is not the issue. Whether welfare, by its availability, has encouraged some girls to choose to become single parents rather than to remain in school and to build an employment career, or whether instead, welfare has only made it easier for such girls to live out their chosen life-style as generally incompetent teenage child-parents providing inadequate socialization and life-preparation for their babies is beside the point.* The issue about

---

*Hurlbert (1984) has underlined the immediate results of teenage pregnancies including low birth weight, a variety of infant physical dis-

public welfare is that when it is provided to the unsupervised young, the inadequately socialized, the inadequately trained, the nonmotivated, and the job-unready, it prevents the development of economic self-sufficiency and social responsibility among parents and children. As was pointed out by Charles Murray (1984), although it may be helpful to some who accept AFDC as temporary aid, it is destructive to the life opportunities of many and to the social order and economy of the society that provides it.

THE CONTEMPORARY EUROPEAN
EXPERIENCE WITH POVERTY

Do other countries have residual welfare dependency? If so, are they studying it objectively, or is the subject politicized along ideological axes as in the United States?

In most of the official ministerial assistance and social insurance offices in Western European countries, the official definition of poverty used is, not surprisingly, the same one that is used in U.S. welfare administration—a monetary definition. Many of the ministries are beginning to become concerned with the concepts of transmitted deprivation (the term used in England) and persistent poverty (the term used by the Poverty Research Committee of the European Common Market—see Schaber 1980). Also, like their U.S. counterparts, Western European social workers, Marxist and neo-Marxist scholars, and other conflict theorists tend to use the relative deprivation definition of poverty, which fits with the income equalization theory as the objective of welfare and social insurance.

Other scholars in Europe, notably Van Doorn (1978) and Seewandono (1984) of the Netherlands and Strang (1970) of Germany have utilized the interactional definition of poverty. Strang describes three

---

orders, loss of educational continuance for the mother-child, low employability prospects for the mother, nonformation of complete families, and continued fatherless pregnancies. More than half of the women become dependent on AFDC and other welfare aid. Most of these children are reared in urban areas in which crime, delinquency, drug addiction, and other social ills predominate. Among the ills associated with teenage, single-parent motherhood are the following added risks for children: higher frequency of admission to hospitals for serious accidents and gastrointestinal infections, poisoning, burns, superficial injuries, and lacerations. These were found to be caused primarily by maternal inexperience, ignorance of dangers, and inability of the mothers to understand the need for adequate child supervision (see Mall 1984).

models of poverty. The first model is related to individual characteristics of people that somehow, and usually through no action on
their part, place them at a socioeconomic disadvantage and relegate
them and their families to a poverty condition. These debilitating
characteristics might include a physical handicap, a severe medical
problem in the family, a legal problem resulting in a severe financial
loss, or some other problem beyond their control. This condition
may be called individual, or as Paul Goodman (1962) describes it in
Growing Up Absurd, "case," poverty.

Another model of poverty, according to Strang (1970), derives
from societal or structural causes. A change in the economy, the
loss of an industry in a town, or a technological or seasonal economic
change can provide the basis of considerable so-called structural
poverty. This is the phenomenon that Paul Goodman (1962) describes
as "class poverty."

Presumably, the first model is the traditional one found in many
preindustrial communities. The "unfortunates" are informally included in all community services and most activities and are usually
protected as full members of the community, despite their handicaps
and deviances. According to Strang, their poverty is accepted by the
community as a substantive urgent need beyond their control. They
are usually provided with in-kind food, housing, clothing, and such.
Thus, in the gemeinschaft type of community, communications and
normative social interactions are maintained between the aided and
the aiding and, therefore, the primary dependency model is neither
alienative nor transgenerational.

The second model of poverty is structurally oriented, based on
the state of a local or national economy. In earlier communities,
the second model was viewed as a universal problem affecting almost
everyone in the community. Because it was a matter of pure chance
that one sector of the laboring population rather than another was out
of work, the social distance between the still-employed and the structurally unemployed was never very great. This type of poverty usually disappears when employment again becomes available.

A third model of poverty, according to Strang, has developed
with the secularization and bureaucratization of assistance. This
model, unlike the first two, tends to increase in number and in depth
over time, and its numbers and depths are unrelated to the state of
the economy and the availability of employment. In the third model,
the behaviors of the chronic welfare clients and the policies and administration of the welfare system interact to promote the continuance
and growth of welfare dependency.

An examination of the welfare and social insurance policies in
Western Europe indicate that welfare dependency is an interactive
product of these systems as well.

Sweden

It should be noted that many Western nations have set out a deliberate plan of welfare-state development, providing welfare and income transfer as well as designs and directives for public education, community affairs, employment, occupational training, care of the sick and aged, housing, and other realms of life reaching to all levels of society. Among these nations, Sweden is best known for its comprehensive welfare-state design. Rydenfelt (1981) indicates that the Swedish economy from 1930 to 1965 was able to install welfare-state mechanisms because taxes were still low and the economic markets outside the welfare-state sectors still functioned well. From 1965 to 1975, the Swedish free-enterprise system was still strong enough to endure increased state taxes and nationalization, but with decreasing growth. Since 1975, however, higher levels of taxes and state controls have resulted in the extinction of larger free enterprises, resulting in growing unemployment and productivity disturbances.

A number of social problems have also arisen in Sweden. Because of increased aspirations for wealth, both parents are employed in most families. Children have less attention from parents, and school progress has been affected. Children in Sweden are now more difficult to teach and control in school than they were prior to the welfare state ethos of child protection. In 1980, 10,000 of the 90,000 school staff left the school system, and 20,000 others have retired early. More than 18,000 substitute teachers (primarily college students) have been hired. Every tenth teacher in Stockholm was reported to have been beaten, attacked, or threatened. Corporal punishment or humiliation of children at home or at school have been prohibited, and social control of children has been diluted. With this has come an increased leniency in the treatment of delinquents and criminals. With the ready availability of income transfer of various types, there has been an increase in youth unemployment and unemployability, a dilution of the Protestant ethic, and a growth in the population of ex-offenders, alcoholics, drug addicts, and other "social outcasts unable to enter or reenter the labor market" (Rydenfelt 1981, pp. 41-44).

The universal health system used 12 percent of the gross national product in Sweden in 1978, and the Swedish medical system reported recurrent emergencies because of low productivity, long waiting lists, and impersonal treatment of patients. Rent control and tenant organizations supported by government sanctions have resulted in a leveling off of new housing. The Swedish welfare-state mechanisms have absorbed almost 12 percent of the GNP, with much of the deficits covered by external borrowing. With the erosion of benefits through inflation and with a growing concern over the government's insecure economic position, the sense of social security originally intended for all Swedish residents has proved quite hollow.

The welfare state of Sweden has a pervasive influence. Under the program, the costs of churches and the salaries of religious functionaries have been taken over by the welfare state. In the process, there has been an erosion of local and voluntary mutual-aid activities formerly under the leadership of church and philanthropic groups. State mechanisms for the aged and poor have grown, but family and community services have given way to the principle of state solidarity. Individuals have been relieved of their obligations for family members, and according to Heckscher (1984), a new moralism has taken over. Rydenfelt (1981) describes the services of the state in its care of the aged, sick, and poor as a form of hired love, which is neither caring nor effective. Huntford (1972) describes Sweden as a "totalitarian state masquerading as a democracy," and indicates that the Swedish experiment has led to an overreliance on an executive bureaucracy that has proved ineffective in providing its purported "womb-to-tomb" services for all. In the process, however, it has increased the size of the non- and less-productive population in various ways. As much as 25 percent of Sweden's labor force now works only part time (Painton 1981). Large sectors of young adults are unemployed and unemployable. The new generation prefers the soft option of a job in the state bureaucracy rather than in productive work. Generous sickness benefits have resulted in a high rate of absenteeism. The productive labor force has been steadily shrinking, as has the Lutheran version of the Protestant work ethic. The social-insurance system (see International Labour Office 1985) suffered a 26 percent deficit covered by government subsidy, over and above a KR27.5 million expenditure for financial aid. The weakness in the Swedish welfare-state plan is apparently that it did not take into consideration the subsidy effect on its population, which has generally weakened the interest of people in employment. Among the Western nations, it is the only one to have shown a negative growth (-2.0 percent) in 1977. Its social expenditures have grown from 12.2 percent in 1963 to 30.7 percent of the GNP in 1977 (see Halberstadt, Haveman, Wolfe, and Goudswaard 1983). The Swedish state has a growing population depending upon it and has less people depending on themselves. It is fast becoming less able to support the load.

The United Kingdom

Social security in the United Kingdom is set out as a system that "seeks to set a minimum standard of living for families and individuals in a time of need" (see <u>Social Security in Britain</u> 1977, p. 1). This consists of a family income-supplement program, seven employment benefit and supplement programs, eight disability and health programs, four widow-aid programs, six industrial-injury programs, three war-benefit and criminal-injury programs, three maternity programs, four children's aid programs, five educational support programs, four health-aid programs, and five housing-aid programs. This is in addition to the National Health Service and other health and welfare institutions (see Cooper 1980-81).

Although the social-insurance programs were initially established on the basis of earnings-related contributions, additional components were added, many as so-called social supports, without offsetting contributory prepayments. These included family allowances, child-aid benefits, allowances for the disabled, and supplementary benefits for which an amended means test was attached. Programs for rental allowances; milk and vitamin supplements; legal aid; and exemptions for health service copayments for dental work, spectacles, and prescriptions were added. All services and benefits were made a right and no requirements were put on recipients to rehabilitate themselves or to seek to become self-sufficient.

By 1978 Peter Jenkins reported that the United Kingdom was "a nation on the skids," with a markedly inferior growth rate, a falling share of manufactured goods in world trade, a slowing productivity rate, and rising labor costs (Jenkins 1978). Tyrell (1978), in his report <u>The Future That Doesn't Work: Social Democracy's Failures in Britain</u>, describes how democracy in the welfare state forces the government to commit itself to expanded benefits without allowing for countermeasures to stabilize the economy. The result, he says, is that "democracy has itself by its tail and is eating itself up fast" (p. 15). Izbieki (1976) reports a growing level of school truancy tied to an increasing number of illegal part-time jobs for children under age 15, with a declining level of educational achievement and later employability. Izbieki reports a growing level of unemployment and unemployability among youth. Wright's report on problem families in Sheffield indicates a high level of dependency on the dole (Wright 1955). Parry, Wright, and Lunn (1967) report a high level of "social failure" among Sheffield families who are themselves products of families that suffered from unemployability and dependence on the dole. Forman (1982) provides us with a detailed account of a welfare-dependent unmarried mother in a high-rise public housing project whose "learned helplessness" was aided and abetted by welfare-state mechanisms. The problem of families intergenerationally locked into the welfare-dependency trap has been described by Townsend (1979) as "transmitted deprivation."

The problem of unskilled and untrained young unemployed workers is discussed by Kramer (1983). She reports that the welfare state

in the United Kingdom is financed at a rate of £17.5 billion, derived
from North Sea oil. What will happen to these unemployed workers
when the oil runs out is a serious issue. The worker who receives
£25 a week on unemployment, plus another £15 for his wife and another
£20 in other benefits is hardly likely to accept a £25 apprenticeship
as an alternative. The atmosphere of ready acceptance of the dole is
indicated by Dunn (1967) in her best-selling book, Poor Cow. Here,
the heroine says about men, "I'll use them if I feel like it, but in the
morning I'll tell them to move on. . . . They like to think you need
them to earn your bread and pay your rent but it's a big lie when you
can perfectly well get it off the assistance" (pp. 18-19).

Welfare dependency in the United Kingdom is closely related to
other social ills. West (1984) confirmed the existence of a delinquent
life-style for youths closely related to chronic unemployment, unemployability, and life on public assistance in multiproblem families.

Problems have appeared in the National Health Service as well.
Despite the availability of health care without cost, 3.5 million people
carry their own medical insurance because of inaccessibility and quality concerns. This has resulted in two parallel health systems, one
for the rich and middle class, and one for others (see King 1981).

Halberstadt, Haveman, Wolfe, and Goudswaard report problems
in the British disability programs. Many individuals have income-
replacement rates beyond the intended 70 percent of regular employment income because of the ability to accumulate benefits from more
than one program. At the same time, there is a large level of hori-
zontal inequity in the system. Some 16 programs are available to the
disabled, and thus, some receive benefits much higher than intended,
while others suffer under the hardship allowances.

The social-insurance costs of £28,500 million have been offset
by a 19 percent contribution by employees and a 33 percent contribution by employers. Tax subsidies provide the 48 percent deficit, in
addition to its provision of £5,500 million of public assistance (International Labour Office 1985). The latter rate has grown at a rate of
about 10 percent per year.

Welfare costs and the National Health Service account for 40 percent of U.K. government spending, up a third from expenditures in
the 1970s. Here again, the United Kingdom has a growing number of
people who are losing their ability to support themselves, and the burden is fast becoming intolerable for the nation to carry.

The Netherlands

The Netherlands has long prided itself as being one of the most
generous states in Europe. Until 1965 aid to persons in need was ad-

ministered by voluntary organizations under federal subsidy. The principal organizations were the Catholic charities, the Protestant social services, and the aid program of the Socialist party. By 1965 the National Assistance Act was passed. Under the act all Hollanders who were themselves unable to cover the costs of life necessities were, in principle, assured of a minimum income (see Sociaal en Cultureel Planbureau 1980). Under the prior law, each of the voluntary agencies was required to use an individualized approach with each client and to provide temporary assistance until self-sufficiency resumed. Under the new law, the poor were viewed as members of a particular interest group and now had a right to apply for benefits to which they were entitled. Each individual could choose a voluntary agency or interest group. The benefits were calculated on the basis of the national per capita income, and the level of subsidy for individuals was based on 40 percent of the national per capita income. An extra weighting formula was introduced for calculating additional support for the aged, the sick, and larger family groups. Almost every form of social insurance was provided, including old-age pensions, pensions for widows and orphans, children's allowances, unemployment benefits, employee's disability benefits, general disability benefits, health insurance, exceptional medical expenses, and special programs for civil servants. Public assistance included supplementary unemployment aid, social assistance, supplementary employment, and housing allowances. The social insurances grew from 10.2 percent of the GNP in 1965 to 23.5 percent of the GNP in 1981. Public-assistance outlays grew from 1.1 percent of the GNP in 1965 to 5.2 percent in 1981. The number of people receiving benefits grew from about 1.5 million in 1965 to over 3 million in 1980. In 1977 (see Halberstadt, Haveman, Wolfe, and Goudswaard 1983) total social insurance expenditures represented f28.5 million, but over 18 percent was derived from federal subsidies.

Of particular concern (see Tifflin 1984) are the disablement programs, where one in every six people in the Dutch work force is on the permanently disabled list, receiving up to 80 percent of former wages. Many of the disabled are also employed in the underground labor market. The average sickness benefit amounted to 90.6 percent of modal earnings in 1981. Tifflin reports that in 1984, 41 percent of the Dutch work force was not employed yet was receiving sizable benefits.

The problems of the Dutch welfare state are compounded by the presence of a foreign work force (200,000 guest workers), many of whom learn to join the subsidized population in the collection of benefits for themselves and their families.

Tifflin (1984) describes Halberstadt's view of the Dutch welfare state. Originally, Halberstadt and other designers of the program had little concern about people taking advantage of the system because they had confidence in the Protestant work ethic of the Dutch people. What the designers had not taken into consideration was that a generation of young people would emerge who would think it quite normal to live on a benefit instead of working. Halberstadt is particularly concerned about the excessive number of disability claims based on psychosomatic symptoms, many of which are fraudulent.

Part of the inflationary effect of the welfare-state expense derives from the built-in cost-of-living increases for public employees and benefit recipients. Another inflationary pressure derives from the rising costs of health care, reported by Simanis and Coleman (1980) as matching that of U.S. health costs. The Dutch health-cost inflation is compounded by the institutionalized nature of the Dutch health system. The various health institutions, organized on a membership basis related to interest groups, are permitted to draw on the government for reimbursement for benefits and services delivered. This provides these institutions with a blank-check arrangement similar to the fee-for-service arrangement between U.S. hospitals and Medicare in the 1970s and early 1980s.

Still another heavy expenditure of the Dutch welfare state is its expenditure on private schools. Van Doorn (1978) notes the growth of school subventions by the state. In the past century such subventions were set at 20 percent; by 1945 they were set at 70 percent. By the 1980s these subsidies reached 100 percent. These subsidies now include not only schools but psychiatric sanitoriums, hospitals, institutions for the retarded, homes for neglected or disturbed children, old-age homes, community health and counseling centers, and other specialized facilities.

With the growth of interest groups, sectarian and secular institutions, service-distributing agencies, and associational pressure groups based on religious or ideological differences, there has developed an attenuation of concern for the individual, and there has developed an increased interest in "pillarization," leading to what Van Doorn describes as "democracy by demonstration." As the pillars or interested groups have grown, they have also become more centralized and more distant from the clients and recipients. Thus, these agencies, which deliver 95 percent of the health and social goods, have become impersonal and less able to restore their clients to self-sufficiency, even if they and the clients were so motivated. These pillars compete for members and centralized dominance. Their emphasis is on client retention rather than on client independence.

It should be noted that many of the expanded benefits were originally planned as a result of the Dutch North Sea natural gas sales. Unfortunately, the price of natural gas has fallen with the recent oil glut, and the reserves are now proving to be much less available than originally estimated. Thus, the Dutch welfare state is beset with increasing demands from a less-productive population and organized interest groups at a time when it can depend on less income and considerable economic problems.

The Federal Republic of Germany

Aid to the poor was in the hands of the church and private welfare agencies in Germany until 1924, when the first government welfare programs were established. In 1962 this program was repealed in the Federal Republic of Germany (FRG), and the Federal Social Insurance Act was substituted. The new act set up three approaches to those in need.

1. A contributory social insurance program including protection in the following areas: health care, invalidity, unemployment, sickness, aged care and retirement, survivors' benefits, and against other life risks;
2. A noncontributory system for civil servants, armed services, war victims, and veterans;
3. A social-assistance program (noncontributory) involving children's allowances, allowances for single-parent families, housing allowances, and general welfare assistance (see Ross and Zacker 1983).

Under the act, subsistence assistance was granted to persons without sufficient funds for life necessities, based on guidelines set up by each "land" (an entity similar to states in the United States and cantons in Switzerland). The subsistence level was related to general cost of living, size of families, and ages of family members. In exceptional cases, the law provides for restricting assistance to basic necessities in the event that a client behaves in an irresponsible manner or deliberately gives up or refuses employment (Ross and Zacker 1983). Another special provision in the act authorizes special assistance such as loans to set up a family business or for preventive medical measures; integration assistance, or special care for the aged or handicapped; and special provisions for drug addicts or discharged criminals. A basic principle of the social-assistance program is that it was designed to serve as a subsidiary to the person's own earnings or income or to the person's social-insurance benefits. The obligation of close relatives is written into the act. Social-assistance programs are specifically prohibited from providing aid as a substitute for self-maintenance. Nevertheless, most grants are based primarily on specified formulas without exercise of welfare-worker discretion in regard to denial of aid because of lack of client effort for self-support. From conferences with government authorities, it is apparent that very little (or no) individuation occurs in the administration of most welfare grants.

Liebfried (1978) indicates that the nature of West German poverty has changed in recent decades: "There is, in general, no poverty in the FRG due to (structural production problems). Rather there is only socially induced poverty . . . which is incrementally (developed) by the system of social security" (p. 247), which is apparently a form of "interactive poverty" brought about by interaction of the affected populations and the welfare industry. According to Liebfried (who supports a higher minimum grant for all), individuation of welfare management according to the client's particular needs and requirements for rehabilitation has been dropped, even though it was the basic principle of welfare reform in the Weimar Republic. Although individuation is still given much lip service in West German social work circles, the welfare establishment is fearful of the consequences of such a policy because it might be utilized as a means of "coercion" of clients to move toward self-support. Thus, the West German welfare establishment has pressed for "a minimum (unearned) pension for everyone," particularly single-female parents, as their "wage for child rearing" (pp. 248-49). According to Liebfried, most welfare-dependents in West Germany, aside from homeless vagrants, ex-convicts, and gypsies, are the products of the "newer policy" based on state subsidy and "relative definitions of poverty." This policy has apparently been a factor in the increased divorce, desertion, and unmarried-motherhood rate. A publication of the Ministry for Family, Youth and Health, Treffpunkt (June 1980), featured the problem of absent fathers. In translation, the article was titled, "When the Father Doesn't Pay, Then Father State Pays." The article made the point that the state as surrogate father provides only money and none of the child-rearing guidance needed by all children. The conclusion one draws from the article is that the state is a very inadequate father.

West German welfare-state problems are compounded by the continued presence and influx of foreign workers. Many workers from non-European Economic Community (EEC) countries arrived in the years prior to the oil crisis of 1974, and most have remained. In addition, EEC (Common Market) policy requires the free movement of goods and labor among the member nations, which now include Greece, Spain, and Portugal. In 1978 there were about two million foreign workers and dependents, including 28 percent from Turkey, 19 percent from Yugoslavia, and 32 percent from EEC countries. There are also thousands of illegal immigrants who arrive in East Berlin by Soviet and Cominform Airlines "on excursions," entering West Berlin and the Federal Republic of Germany without visas under the FRG's United Germany policy. Many of these visitors remain under FRG asylum policies. Many of the immigrants, legal and illegal, eventually become eligible for social insurance and assistance benefits and become dependents. A sizable Turkish community has developed in various West German cities, along with Islamic centers, separate religious schools, and considerable social conflict between natives and newcomers. Much of the growth in drug offenses and other crimes has been attributed to the Turkish population. Many of the jobs refused by West Germans in their preference for welfare grants have been taken on by foreign workers. The isolationist patterns of the Turkish immigrants and the problems of their assimilation and acculturation have been described as a "social time-bomb" waiting to occur.

The Treffpunkt report listed 180,000 single parents (primarily women) with children under six supported by welfare-state programs in 1980. About seventy-three thousand children were provided for in that year. State subsidy of these children has created a condition comparable to the situation in the United States, where parental nonsupport has been made possible by welfare availability. Parents are then free to have more children with other partners.

Another problem of the FRG welfare state was brought about by the courts, which ruled in 1982 that every citizen has a constitutional right to an overseas vacation, even while receiving unemployment benefits. Thus, a West German worker receiving 68 percent of his or her regular wage on unemployment compensation for one year can choose to spend it at a foreign resort, thus becoming unavailable for other employment. That raises the cost of unemployment compensation, which is actuarily calculated as an alternative to unemployment caused by structural factors. Two million people in the Federal Republic (7 percent) were on unemployment compensation in 1982 (see Trimborn 1982). After a year on unemployment compensation, unemployed workers are eligible for welfare at 58 percent of their net wage, plus extras based on size and situation of their families.

West German welfare-state regulations are so complex that a simplified pamphlet listing a person's rights comes to 300 pages (see Painton 1981). Extensively generous loopholes maintain vast programs that tend to perpetuate the interaction between the welfare clientele and the welfare-state system.

Because of the expense and social conflict brought about by foreign workers, the West German government recently offered a lump-sum payment of $4,200 for each foreign resident and $600 for each child, plus reimbursement of all social security premiums paid if the foreigners would leave West Germany (see Smarz 1983). It is yet to be seen whether this will be effective, in view of the "revolving-door" condition of immigrant control in West Germany.

Smarz cites the generosity of the West German welfare state. In the case of a middle-class couple with an 18-month-old daughter, $6,500 was given in allowances. In addition 12 free prenatal medical examinations, $2,500 in state-provided delivery, and substitute wage payments to the mother during six months' pregnancy leave were all provided. The family is also eligible for $28 per month preschool costs and free education through college or graduate school. If the daughter is unemployed on completion of her education, she would receive $140 per month plus her rent while she is looking for a job.

As a result of the high tax rates occasioned by the welfare state, a sizable tax-evading underground economy has developed. Craftsmen were reported by Painton (1981) to have earned $17 billion in unreported jobs in 1978. Entire housing developments have been built with "black labor." Smarz estimates that the "black economy" has now become respectable, with kickbacks and bartering of services now quite common. There is considerable underground employment, while beneficiaries collect unemployment benefits and welfare. Many items are cheaper when paid for in cash.

Schmidt (1982) questions the presence of poverty in the FRG if the national per capita income is DM18,769. (At the time, that amounted to a per capita income of $7,883.) Despite this, with a welfare grant of about $774 per month in a family of four (aside from special grants), a conference of welfare clients in West Germany heard complaints that families could not afford hairdressing services for daughters and soccer game admissions for their sons.

Halberstadt, Haveman, Goudswaard, and Wolfe (1983) report that the West German programs for the disabled are exemplary in that they emphasize rehabilitation, retraining, and gainful reemployment. This is not the case with the other FRG programs (Liebfried 1978).

Strang (1984) reports that as the federal support of income programs has recently been restricted, a greater load has been placed on local governmental units, especially in the larger cities, where higher unemployment rates are becoming common; large numbers of single women head families with children; and alcoholics, drug addicts, and young "dropouts" congregate. When federal financial assistance is given, approximately 60 percent of this goes for "aid for care" or "special assistance," but expenditures to "help people to help themselves" tend to be entirely absent (pp. 14-15). There is a heavy reliance on institutions for care of family members, despite the fact that they can be served at home for one-seventh of the cost. The welfare state avoids pressing the retention of the elderly and sick by grown children and extended family at home.

Strang also notes a heavy concentration of social-assistance cases and higher grants per family in the northern provinces. Strang attributes this not to a difference in clientele, but to differences in administrative policies in the south, where self-care, self-maintenance, family responsibility, and self-reliance are emphasized. He also notes that assistance rendered by lay persons promotes "social solidarity" in people who might otherwise become isolated. Contact by volunteer workers with clients would be more likely to provide an effective learning model for someone to move toward self-dependence. He indicates that this was less likely in contacts with professional welfare workers, who tend to view dependency on the state as normative.

Strang indicates that the principle of helping the client to help himself has been submerged into income maintenance by increased appeals to the courts by clients and by the reluctance of welfare workers to interfere with their client's life-style. The so-called welfare self-help groups are, by and large, led by women clients who are themselves heads of dependent welfare households, according to Strang. Thus, the emphasis of these welfare self-help organizations is placed on increased welfare aid rather than for mutual aid and self-reliance.

Strang finds that the West German social-assistance system serves to socialize the clients, but in the direction of client dependence rather than toward self-reliance. The social state promotes a here-and-now mentality rather than a planful approach to life. The slogan of the welfare clientele is, He who doesn't shirk is a jerk (meaning "crazy"). Strang describes how the process of welfarization develops among the West German clientele. Primary welfarization occurs as a consequence of objective and acute conditions of need, but secondary welfarization occurs when dependence upon the system turns into a lasting condition, thus locking the client into "welfare-as-a-way-of-life." Secondary welfarization occurs when the welfare system no longer promotes self-help and when the client no longer perceives any alternative to dependence on welfare. It is furthered by criteria that reward dependency, attracting a population that is gratified by such rewards. Secondary welfarization is promoted by inadequate and inappropriate socialization of children, by inadequate effectiveness of children's education for employment at home and in school, and by inadequate public concern about unnecessarily extended reliance on welfare. According to Strang, secondary welfarization in the FRG is predominantly a matter of divorced women with children (rather than unmarried mothers, as in the United States). The socialization by welfare mothers of their children promotes intergenerational welfare dependency.

Strang found that financial aid alone is not sufficient to solve welfare-family problems. An adequate and specialized professionalism is required for families to prevent or resolve welfarization. This special type of professionalism requires workers who are sensitized to promote self-help. Strang suggests that financial assistance

be converted into repayable loans to encourage earlier self-reliance
and self-responsibility. He also suggests that the grants given families be graded up or down as needed to encourage self-help; that a
purposeful public relations program for self-help, prevention of dependency, and revision of stigma attitudes be designed and promoted;
and that special attention be given to families with children to ensure
that the children are socialized toward self-dependency.

The West German social-insurance program cost the West German government about DM107,000 million in 1980 (see International
Labour Office 1985). This is over and above employer and employee
contributions (which cover 70 percent of the costs). In addition, the
government paid out DM15,200 million in financial assistance, making
a total of DM122,200 million in noncontributory benefits. Halberstadt,
Haveman, Wolfe, and Goudswaard (1983) indicate that the 1977 expenditures for the welfare state represented 26.5 percent of the GNP,
an increase from 16.9 percent in 1963. Productivity growth held up
at 2.7 percent in 1971 but fell to 0.9 percent in 1980. Labor-force
participation has fallen for older males (over 54 years), apparently
because of the availability of retirement and other benefits.

Although the West German welfare-state program has promoted
the growth of a welfare-dependent population, the economy has held
up through 1980. It is questionable, however, how long West Germany
will be able to continue its policies at their present level as productivity gains level off. The difficulty of cutting back once a grant has
been made available is as much of a problem in the Federal Republic
as it is elsewhere.

France

The development and structure of social security in France is
unlike that of any other Western nation, according to the Fondation
pour la Recherche Sociale. It developed much later than in other
countries. It is also different in that it is organized through a variety
of occupational associations making up a mosaic of coverage. There
are three streams of coverage: (1) old age and survivor's insurance;
(2) illness, invalidity, disability, accident, and health insurance; and
(3) family protection. The first stream also provided pensions to
families with many children (even if the parents had never been employed). Under the third stream, assistance to families, orphans,
and "isolated parents" is also provided. Each occupational group
provides the services for each stream.

The complex mosaic of programs combines social insurance
with financial aid in such a manner that no one (administrator, researcher, social worker, or client) really can determine which bene-

fits have been earned under payroll contributions and which are really welfare aid. This complexity has deleterious effects in terms of confusing anyone who might seek to remain self-sufficient or who might seek to promote such an ethic in the society. Each of these institutions is nonprofit and composed of local organizations and units, all supervised by government, but under the different ministries that cover the various occupations and industries. The supervising agencies check the programs for maximum and minimum payments, payment criteria, and to some extent, payment equity. Under French patterns of bureaucracy, it is the letter rather than the spirit of the law that is enforced by the government. Unemployment benefits were organized outside of this mosaic in 1979, under a complex of government-supervised employment associations that provide both special and basic unemployment insurance, fixed-sum benefits, and a guaranteed income after the end of insurance compensation (see Fondation pour la Recherche Sociale 1980).

In addition, there is also a supplemental welfare system (aide sociale) providing medical aid; support for the elderly; support for the handicapped; children's aid; and aid to institutions and homes where children, aged, ill, or handicapped are served. There are also voluntary nonprofit insurances for additional health-care coverage.

Under French law, every economically active person must be covered for social insurance through his or her specified occupational association. The benefits and prepayments vary from occupation to occupation.

Because of the differences in benefits and prepayments among occupational associations, serious questions of equity exist in the French social security system. Also, because some occupations entail less risks than others, the shared-risk principle of social insurance is omitted in the French system. Because some industries wane as others expand and the system is fragmented, there is no sharing of the differential economic risks.

The disability system in France, according to Halberstadt, Haveman, Goudswaard, and Wolfe (1983), serves all handicapped from birth through employment and emphasizes work rather than income assistance. On an average, the program provides 50 percent of the legal minimum wage for salaried workers.

The French social security system is unusual in that it depends on government support for only about 10 percent (approximately Fr68,700 million). The system is heavily dependent upon employee and employer contributions (International Labour Office 1985). In addition, the French government supplies about Fr115,000 million in financial assistance. Walliman (1983) indicates that while the French gross national product increased 138 percent from 1970 to 1977, the social expenditures of the government rose 200 percent. Health expenditures rose even faster (230 percent) in the same period. The required retirement minimum increased some 380 percent (to Fr 13,800 per year) during the same period, although prices had risen over 120 percent during the same period. The deficits encountered in the health services were covered by lessened expenditures in family allowances. With a rapidly aging population requiring ever more health expenditures, it is doubtful that these deficits can continue to be covered by savings in other programs.

French social insurance and assistance costs in 1980 amounted to 25.5 percent of the GNP, up from 22.7 percent in 1975. Walliman predicts a growing resistance to employer-participation contributions for social insurance. This resistance may occur in the form of more capital-intensive (and fewer labor-intensive) operations. Walliman also predicts serious conflict between industrywide worker groups and employers as unemployment grows. The social-insurance occupational associations, which were originally built on employer-employee cooperation, may crack under the economic strain of rising labor costs and unemployment rates.

Walliman also reports on the inefficiency of the social services, which serve primarily as disbursers of social benefits and financial aid and which have become increasingly specialized, bureaucratized, and enlarged in number. He reports a pervasive development of social work bureaucratic momentum and a concurrent maintenance of dependency among the clientele. The system is highly centralized, and decentralization and democratization are strongly resisted. Rosa (1982) and Cook (1982) are concerned with the rising costs of the social-insurance system and in the lost gross national product brought on by the 39-hour week, the addition of a fifth week of vacation for workers, the lowered retirement age (to 60 years), the 40 percent increase in minimum pension, and the 50 percent increase in family and housing allowances for low-income families (enacted in 1982). There was also a further increase of 2.6 percent in the basic minimum wage and other allowances. Employers' social-insurance contributions were among the world's highest prior to the 1982 increase, amounting to 38.40 percent of each worker's total wage. Painton reported in 1981 that the government faced a deficit of US$5.3 billion as a result of welfare-state costs having risen to US$124 billion in 1980 (compared with US$31 billion in 1970).

France's social security and assistance problems are aggravated by thousands of guest workers and their families (Echikson 1983b). In 1978 there were 1,643,000 such guests from Portugal (23 percent), Algeria (22 percent), Spain (11 percent), Morocco (11 percent), Italy (11 percent), Tunisia (5 percent), and other countries (17 percent) (see Conference Board 1980). All residents are now required to carry identity papers, and North Africans must complete a detailed immigration document.

The high social security contributions have led to a substantial growth in France's underground economy. The "black" firms carry no social insurance and enjoy a strong profit and price advantage in hiring moonlighters and illegal immigrants. This leads to high unemployment and other costs. As the "black" labor force grows, the social system is weakened (Echikson 1983a). Castaing (1975), Benoit (1975), and Drouin (1975) report that there are many people still in poverty. Many young people, vagrants, and gypsies are not in the labor force, and there is persistent poverty that is intergenerational. The previously cited foundation report indicates that there are poor people with multidimensional problems and handicaps that keep them dependent, despite or because of, the variety of cash benefits available to them. For many of these poor, it would be extremely difficult to take advantage of advancement opportunities. The foundation describes this problem as a "vicious cycle," in that untrained, ill-schooled, welfare-dependent people find it difficult to escape dependency or to know how to train their children to do so. It also reports that the relationship between French social workers and the poor shows a one-sided alignment of financial assistance and a bilateral alignment of mutual dependence. The poor receive financial help and the social workers get continued dependent behavior, which justifies their continued almonry services. The foundation also reports that there is a sizable number of single-parent families supported under a variety of allowances and assistances. The nonworking single-parent families are eligible for family allowances, income supplements, family compliments, orphan allowances, health care, special education allowances, rent allowance, personal housing aid, and many others. This makes it conceivable to believe that the sum of the benefits and allowances leaves little room for employment motivation for the parent and, later, for the children.

Thus, it is apparent that the French welfare state, despite its broad population coverage and its sizable social service establishment, has considerable problems with people who are not in the "official" labor force and who are chronically dependent. France has, at great expense, succeeded in seeing to the humane purposes of its welfare system, but in the process, because of its policies and culture (which omit emphasis of employment over welfare dependency), it has endangered its economy and trapped sizable populations in persistent dependency.

Italy

The Italian program of tutto per tutti ("everything for everyone") has been bogged down by bureaucratic red tape and political haggling

to the point that promised services cannot be found and are seldom
available. The Italian health service has underpaid and overworked
doctors and nurses and is experiencing frequent hospital-worker
strikes. What was on paper probably the most perfect health-care
service in the world has in practice been a disaster (see Widman
1983). In Italy, too, the availability of disability pensions has resulted
in a redefinition of the term disability to mean "reduced earning power
capacity," so that currently over five million able-bodied workers are
receiving disability.

An example of how temporary dependency becomes chronic was
reported by Schanche (1982). In 1974 the Singer Sewing Machine Company closed a factory in Leini, Italy, and laid off the workers. The
Italian social-welfare system provided cassa integrazione (special
unemployment insurance designed to tide employees over during plant
reorganization). By 1982 many former Singer workers were still receiving up to 92 percent of their former wages (eight years later) and
were still reported as unemployed. At least 11,000 Italian workers
had been on unemployment compensation for a decade in 1982. In
1967 the social security system had a shortfall of US$417 million.
By 1980 the deficit went to US$6.7 billion, and the deficit for 1985 is
expected to reach US$25 billion. Halberstadt, Goudswaard, and Le-
Blanc (1984) reported that growth of gross domestic product (GDP)
had fallen in Italy from 5.9 percent in 1976 to -0.2 percent in 1981
and 0.8 percent in 1982. Consumer prices had risen from a 3.9 percent average yearly rate of increase for 1961-70 to 19 percent in 1981.
Unemployment had reached 9.9 percent in 1982 after an average of 5
percent from 1961 through 1975. The Italian general government expenses went from an average of 32.4 percent of the GDP for 1961-70
to 52.6 percent of the GDP in 1982. This represents a growth of over
62 percent in 12 years. Italy's net government borrowing exceeded
10 percent of the GNP in 1981. In 1980 the benefit recipients as a
percentage of total employment reached 85 percent, compared with
43 percent in 1963. This is almost a 100 percent increase, and one
questions who is left to produce the taxes to pay for the deficits.

Halberstadt, Haveman, Goudswaard, and Wolfe (1983) report
that the Italian disability programs have been used as a supplement
to unemployment benefits to older workers. This is primarily because
of the more liberal interpretation of eligibility criteria for disablement. Thus, there has been an average growth rate of beneficiaries
of over 8 percent per year.

Castellino (1982) described the Italian social-insurance system
as consisting of a large number of separate programs. Pensions in
1982 represented about 12 percent of the GNP. In 1975 an indexation
revision of benefits actually overindexed the pensions by 12 percent
for every 1 percent increase in prices. As a result, average pensions

in the general retirement program increased from 30 percent to more than 60 percent of the per capita net national income between 1955 and 1980. Thus, in addition to demographic changes (increase of aged population), the crisis in Italian social security is also because of overly generous benefits. Despite increases in the payroll tax rates from 9.0 to 24.2 percent between 1955 and 1980, the contributions are still not sufficient to cover benefits.

According to Castellino, "The Italian system redistributes a great deal of money in a haphazard fashion" (p. 50). This problem has occurred because Italian social policy is based on the view that "every pensioner is a person in need" (p. 51) (rather than that a pensioner should be given benefits based on his past earnings and payments into the social security system). In Castellino's view, the requirement of a 24.2 percent payroll tax with no ceiling is dysfunctional in terms of a stable productive economy. Castellino indicates that the Italian social security policy has a built-in bias that "pushes the pay-as-you-go system (too) far." Italy has also "superimposed a strong dose of Mediterranean lightheartedness. . . . The current deficit is a permanent nightmare for the Treasury minister" (pp. 56–57). He reports that millions of disability pensioners who are not really disabled live in the poorer regions of the country. He indicates that these pensions are really a permanent unemployment dole in disguise, which discourages entry of such persons and their children into the labor market. It should also be noted that the high payroll tax for social security encourages a withdrawal from the open labor market to the underground labor market, for which there are no retirement, sickness, or other programs and from which no taxes are drawn. In Castellino's opinion, the outlook for Italy's economy, its social-insurance system, and its long-run security or humane efforts for the needy are bleak.

Denmark

The problems of the Danish welfare state are, in many ways, similar to those of the Netherlands. Danish labor shifted from the private productive sector into the nonproductive public sector. Public employment numbers rose from 550,000 to 750,000, an increase of 27 percent from 1974 through 1979 (see Painton 1980). Halberstadt, Goudswaard, and LeBlanc (1984) report that government expenditures in 1982 reached 60.6 percent of the GDP, a rise that began in 1971 when it was 44.5 percent of the GDP. The average from 1961 to 1972 was only 33 percent. Thus, the new government expense level in 1982 represented a rise of 84 percent over the level of the period 1961–70. Benefit recipients as a percentage of total employment

reached 64 percent after a rate of 28 percent in 1965 and 41 percent
in 1970. Industrial employment, according to Painton (1980) went
from a high of 420,000 workers in 1974 to 380,000 in 1979. Thus, the
nation's expenses expanded, but the tax base to pay for it declined.
The welfare benefits remained high with some "double-dipping" in the
welfare-state programs. This was characteristic of workers who limited their work hours in order to collect partial unemployment compensation for working only part-time, while working on the "black"
during the officially listed unemployed time. Stovall (1980) reports
that the Danish welfare state offered full medical care, full education,
child allowances, rent subsidies, aged and survivors' pensions, as
well as maintenance of previous living standards for the unemployed.
He quotes social reformer N. F. S. Gruntwig as saying, "Few have
too much and even fewer too little" (p. 15). Stovall reports that by
early 1980 the Danish welfare state was consuming 52 percent of the
gross national product. Denmark's heavy borrowing to support its
social programs brought its zero trade deficit of 1970 to a debt of
Kr20 billion by the end of 1979.

Political leaders elected as a result of the tax revolt of the
1970s sought to curtail excessive welfare benefits, but this did not
succeed by 1980, primarily because of labor union opposition. Stovall
also discusses proposals for refusal of pensions to the well-to-do
(after they had paid in a lifetime of pension contributions at rates
slanted to provide larger pensions for lower incomes). This proposal
was apparently also unacceptable to the electorate.

Marshall (1984) explains that currently the Gilstrup tax revolt
has waned and that the Danish welfare state remains one of the most
comprehensive welfare systems in Western Europe without the tax
base needed to pay for it.

Belgium and Ireland

Similar problems of expansive government spending related to
welfare-state generosity, along with low economic growth, high unemployment rates, and inadequate tax bases to pay for them, have arisen
in Belgium and Ireland (see Halberstadt, Goudswaard, and LeBlanc
1984). In Belgium, unemployment recently reached 13.9 percent, and
in Ireland, it reached 12.1 percent in 1982.

Hirsch's research on the security state (1980) found that although
one may buy security from the state, the price is often the loss of personal spontaneity for the individual and productivity for society.

The situation of the Western European welfare states is remi-
niscent of the problems reportedly faced by a grand duke in one of the
isolated regions in Europe. He was full of goodwill for his common-

ers and was quite generous to them. When families came to him for help because they could not make ends meet to support their children, he would give them a monthly allowance—but he was upset that they spent his money unwisely, conceived even more children, and frequently worked less than they could. He could not let the children starve and so he increased his allowances to these families. Many of the youth in his duchy dropped out of school or did badly in school and ignored the vocational training programs that he sponsored. Instead, they would come to him for help—and he could not let them starve, even if their pride prevented them from taking the available jobs of cleaning, serving, cooking, collecting garbage, and so forth. So, the duke gave them a monthly allotment and, instead, hired foreigners to do these jobs. He also had to take care of the elderly, the sick, and the handicapped in his duchy.

One day, the duke sat down in his office to see how much money would be needed for the week. He suddenly realized that so many people were getting allowances from him that he was collecting less and less taxes each month. The banks in his duchy also complained to him that fewer people were saving their money, probably because they counted on the duke to help them if they ever were in financial trouble or needed money in their old age. The factory managers complained that they had to import foreigners to keep up production and now even some of the foreigners were calling on the duke rather than working. The local police complained to him that many idle people got into trouble more often than when they were busy working or in school. Even the people he was helping regularly complained because they wanted more and higher allowances.

Finally, the duke decided that he had had enough. He called on his council of advisers and asked them to study the problem and to tell him what or who was causing all this trouble for him. After a week of study and discussion, they came back with their report. They had found the culprit—it was the duke and his simplistic generosity. As long as the duke assured people that he would always take care of them, they no longer thought twice about spending money, having more children, saving for their old age or illness, or even making sure that their children learned a trade to support themselves.

Instead of being so generous to everyone, the advisers recommended that the duke give some limited help only to the really aged and sick. To all others, he would have to say, "Go to work." To the guest workers, he would have to say, "Please go home to your own country and try to work out your problems there—we have our own problems here." To parents with children, he would have to say, "I will give you only a bit of food for the children for a month— after that you'll have to earn your own bread." To the adults in the community, they recommended he say, "You will have to save for your own retirement—if you can't do that, I'll collect from you every month and I'll return the money when you're old." The duke finally realized that for his people to learn to take care of themselves, he would have to keep from interfering with their learning. The good duke learned that it is more humane to let people learn to deal with life on their own.

What the duke learned, the welfare states need to know. Just enough short-term help to help a person to help himself or herself is more useful and humane than to provide continuing substantive support. It is too much help when the helped person stops trying to help himself or herself.

REFERENCES

Aldous, Joan, and Reuben Hill. 1969. "Breaking the Poverty Cycle: Strategic Points of Intervention." Social Work 14:3-12.

Anderson, C. LeRoy. 1965. "Development of an Objective Measure of Orientation toward Public Dependence." Social Forces 44: 107-12.

Anderson, Martin. 1978. Welfare: The Political Economy of Welfare Reform in the U.S. Stanford, Calif.: Hoover Institution.

Auletta, Ken. 1982. The Underclass. New York: Random House.

Axinn, June, and Herman Levin. 1975. Social Welfare: A History of the American Response to Need. New York: Dodd, Mead.

Bane, Mary Jo, and David T. Ellwood. 1983. The Dynamics of Dependence: The Routes to Self-Sufficiency. Cambridge, Mass.: Urban Systems Research & Engineering.

Banfield, Edward C. 1969. "Welfare: A Crisis without Solutions." Public Interest 16:89-101.

_____. 1968. The Unheavenly City: The Nature and Future of Our Urban Crisis. Boston: Little, Brown, p. 211.

Battle, Esther S., and Julian B. Rattner. 1963. "Children's Feelings of Personal Control as Related to Social Class and Ethnic Group." Journal of Personality 31:482-90.

Bedeau, Hugo Adam. 1967. "Egalitarianism and the Idea of Equality." In Equality, edited by Roland Pennock and John W. Chapman, pp. 3-27. New York: Atherton.

Beeghley, Leonard. 1978. Social Stratification in America: A Critical Analysis of Theory and Research. Santa Monica, Calif.: Goodyear.

Benoit, Jean. 1975. "Welfare Outlaw: France's New Poor." Manchester Guardian Weekly, vol. 113, no. 17 (October 26).

Berger, Joseph. 1984. "Rise in Illegitimate Children Worries New York Officials." Fresno Bee, August 16, p. 11.

Butler, R., and J. Heckman. 1976. "The Impact of the Government on the Labor Market Status of Black Americans: A Critical Review of the Literature and Some New Evidence." Unpublished manuscript received by Ralph Segalman, November 1976.

Castaing, Michel. 1975. "France's New Poor: A Fight for Survival." Manchester Guardian Weekly, vol. 113, no. 16 (October 19).

Castellino, Onorato. 1982. "Italy." In The World Crisis in Social Security, edited by Jean-Jacques Rosa, Richard Hemming, John A. Kay, Oronato Castellino, Noriyuki Takayama, Ingemar Staht, Martin C. Janssen, Heinz H. Müller, Sherwin Rosen, Karl Heinz Jüttemeir, and Hans-Georg Petersen, pp. 48-69. San Francisco: Institute for Contemporary Studies.

Chapman, William. 1977. "The Welfare Enigma: Despite All the Programs, Reforms, and Billions, the Poor and Their Problems Will Not Go Away." Manchester Guardian, June 5, pp. 1, 17-18.

Cherlin, A. J. 1977. Work, Welfare, and the Increase of Women Living Independently of Men. Research Project GC-1758. Washington, D.C.: U.S. Department of Labor.

Chilman, Catherine S. 1979. Adolescent Sexuality in a Changing Society: Social and Psychological Perspectives. DHEW Pub. No. NIH 79-1426. Washington, D.C.: U.S. Government Printing Office.

Coll, Blanche D. 1969. Perspectives in Public Welfare. Washington, D.C.: U.S. Government Printing Office.

Conference Board. 1980. International Labor Migration. Economic Road Maps, nos. 1890 and 1891. New York: Conference Board.

Cook, Dan. 1982. "France: Going Up." Los Angeles Times, February 28, pt. 5, p. 3.

Cooper, S. A. 1980-81. Social Security, Welfare, and Benefit Schemes. London: Policy Studies Institute.

Dahrendorf, Rolf. 1958. "Toward a Theory of Social Conflict." Journal of Conflict Resolution 2:170-83.

Doolittle, Frederick, Frank Levy, and Michael Wiseman. 1977. "The Mirage of Welfare Reform." Public Interest 47:62-87.

Dornbusch, Sanford M., J. Merrill Carlsmith, Steven J. Bushwall, Philip L. Ritter, Herbert Leiderman, Albert Hastorf, and Ruth T. Gross. 1985. "Single Parents, Extended Households, and the Control of Adolescents." Child Development 56:326-41.

Drouin, Pierre. 1975. "The Limits of Poverty." Manchester Guardian Weekly, vol. 113, no. 17 (October 26).

Dunn, Nell. 1967. Poor Cow. Garden City, N.Y.: Doubleday.

Durham, Eugene. 1973. "Have the Poor Been Regulated? Toward a Multivariate Understanding of Welfare Growth." Social Service Review 47:339-59.

Echikson, William, Michael R. Meyer, Maks Westerman, and Rita Dallas. 1983a. "Europe's Black Economy." Newsweek, July 25, pp. 12-13.

Echikson, William, Alex Gelber, Ronald Henkoff, Zofia Smardz, and MacLean Gander. 1983b. "The Welfare State in Crisis." Newsweek, July 25, pp. 8-14.

Ellwood, David T., and Mary Jo Bane. 1984. "The Impact of AFDC on Family Structure and Living Arrangements." Research report prepared for the Department of Health and Human Services under grant no. 92A-82.

Ellwood, David T., and Lawrence H. Summers. 1985. "Poverty in America: Is Welfare the Answer or the Problem?" Paper presented at Conference on Poverty and Policy: Retrospect and Prospects at Williamsburg, Va., December 6-8, 1984.

Feagin, Joe R. 1975. Subordinating the Poor: Welfare and American Beliefs. Englewood Cliffs, N.J.: Prentice-Hall.

Ford Foundation. 1980. Current Interests of the Ford Foundation. New York: Ford Foundation.

Ford, Thomas R. 1962. "The Passing of Provincialism." In The Southern Appalachian Region, edited by T. R. Ford. Lexington, Ky.: Thomas R. Ford.

Forman, Rachel Zinober. 1982. Let Us Now Praise Obscure Women: A Comparative Study of Publicly Supported Unmarried Mothers in Government Housing in the U. S. and Britain. Washington, D. C.: University Press of America.

Fondation pour la Recherche Sociale. 1980. Poverty and the Anti-Poverty Policies: The French Report Presented to the Commission of the European Communities. Paris: Fondation pour la Recherche Sociale.

Garland, Patricia. 1967. "Teenage Illegitimacy in Urban Ghettos." In Unmarried Parenthood: Clues to Community Action, pp. 24–39. New York: National Council on Illegitimacy.

Gilder, George F. 1981a. Wealth and Poverty. New York: Basic Books.

_____. 1981b. "Wealth and Poverty." Firing Line. Public Broadcasting System, February 8. Printed version no. 447; Columbia, S. C.: Southern Educational Communications Association.

_____. 1978. Visible Man: A True Story of Post-Racist America. New York: Basic Books.

_____. 1973. Sexual Suicide. New York: Quadrangle Books.

Goodman, Paul. 1962. Growing Up Absurd. New York: Vintage-Random.

Halberstadt, Victor, Kees Gouswaard, and Bart LeBlanc. 1984. "Current Control Problems with Public Expenditure in Five European Countries." In Europe's Stagflation, edited by Michael Emerson. Oxford: Clarendon Press.

Halberstadt, Victor, R. H. Haveman, K. P. Goudswaard, and B. L. Wolfe. 1983. "The Economics of Disability Policy in Selected European Community Countries." In Economie en arbeids-ongeschiktheid, edited by V. Van den Bosch and C. Peterson, pp. 17–66. Antwerp: Kluwer, Deventer.

Halberstadt, Victor, R. H. Haveman, B. L. Wolfe, and K. P. Goudswaard. 1983. "Inefficiencies in Public Transfer Policies in

Western Industrialized Democracies." In Ombuigingen in de socialezekerheid: Ein paradox voor de economie van de jaren-tachtig, edited by G. M. J. Veldkamp, pp. 25-51. Antwerp: Deventer.

Heckscher, Gunnar. 1984. The Welfare State and Beyond: Success and Problems in Scandinavia. Minneapolis: University of Minnesota.

Herzog, Elizabeth. 1963. "Some Assumptions about the Poor." Social Service Review, 37:389-402.

Hill, Martha S., and Michael Ponza. 1983. "Poverty and Welfare Dependence across Generations." Economic Outlook, USA 10: 61-64.

Hirsch, Joachim. 1980. Der Sicherheitsstaat. Frankfurt: Europaische Verlagsaustalt.

Homans, George. 1950. The Human Group. New York: Harcourt, Brace.

Hulbert, Ann. 1984. "Children as Parents." New Republic, September 10, pp. 15-23.

Huntford, Roland. 1972. The New Totalitarians. New York: Stein & Day.

International Labour Office. 1985. The Cost of Social Security: Basic Tables, Eleventh International Inquiry, 1978-1980. Geneva: International Labour Office.

Izbieki, John. 1976. "Dole Gueve 'Breeding Ground for Revolution.'" Daily Telegraph (London), June 1, p. 2.

Jenkins, Peter. 1978. "A Nation on the Skids." Manchester Guardian, October 8, p. 5.

King, Nick. 1981. "Tired of Waiting for Free Health Care, More Britons Are Turning to Private Doctors." Los Angeles Times, May 27, pt. IA, p. 2.

Kluckhorn, Florence. 1958. "Family Diagnosis: Variations in the Basic Values of Family Systems." Social Casework 39:63-85.

Kramer, Jane. 1983. "Letter from Europe." New Yorker, September 5, pp. 94-101.

Kristol, Irving. 1971. "With the Best of Intentions, the Worst of Results." Atlantic Monthly 228:45-47.

Liebfried, Stephan. 1980. "Notes on Mistargeting Welfare Benefits." In Reproduktionrisiken, soziale Bewegungen, und sozial Politik, edited by Stephan Liebfried, pp. 248-49. Bremen: Research Center on Social Movements and Social Policy, University of Bremen Press.

_____. 1978. Public Assistance in the United States and the Federal Republic of Germany: Does Social Democracy Make a Difference? Research Center on Social Movements and Social Policy, Working Paper no. 1. Bremen: University of Bremen Press.

Mall, Janice. 1984. "Children Born to Teen Mothers at Risk." Los Angeles Times, November 18, pt. 6, p. 8.

Marshall, Tyler. 1984. "Funding Increases: Welfare in Europe, A Firm Hold." Los Angeles Times, July 9, pt. I, pp. 1, 8.

Miller, Walter B. 1958. "Lower Class Culture as a Generating Milieu of Gang Delinquency." Journal of Social Issues 14:5-19.

Ministry for Family, Youth and Health. 1980. "Wenn der Vater nicht zahlt . . . dann zalt 'Vater staat.'" Treffpunkt. Bonn: Ministry for Family, Youth and Health, pp. 4-5.

Moore, Kristin A., Sandra L. Hofferth, Steven B. Caldwell, and Linda J. Waite. 1979. Teenage Motherhood: Social and Economic Consequences. Washington, D.C.: Urban Institute, pp. 1146-47.

Morgan, James N. 1978. "Statement." Hearings before the Subcommittee on Public Assistance, Committee on Finance, 95th Cong., 2d sess., S. Rept. 2084, vol. 4, pp. 999-1001.

Mott, Frank. 1983. Welfare Incidence and Welfare Dependency among American Women: A Longitudinal Examination. Worthington, Ohio: Center for Human Resource Research.

Moynihan, Daniel, ed. 1968. On Understanding Poverty. New York: Basic Books.

Murray, Charles. 1984. Losing Ground: American Social Policy, 1950-1980. New York: Basic Books.

Nathan, Richard P. 1978. "Statement." Hearings before the Subcommittee on Public Assistance, Committee on Finance, 95th
   Cong., 2d sess., S. Rept. 2084, vol. 4, pp. 996-99, 1055-57.

Painton, Frederick. 1981. "Reassessing the Welfare State: A Hu-
   manitarian Dream Becomes a Nightmare." Time, January 12,
   pp. 32-33.

_____. 1980. "Woes of the Welfare State." Time, December 1, pp.
   11-15.

Parry, Wilfred, Catherine Wright, and John H. Lunn. 1967. "Sheffield Problem Families: A Follow-up Survey." Medical Offi-
   cers 118, no. 10 (September 8):130-32.

Parsons, Talcott. 1964. The Social System. Glencoe, Ill.: Free
   Press.

Piven, Frances Fox, and Richard A. Cloward. 1971. Regulating
   the Poor. New York: Random House.

Pruitt, Walter A., and John Van de Castle. 1962. "Dependency
   Measures and Welfare Chronicity." Journal of Consulting Psychology 26:559-60.

Rein, Martin, and Lee Rainwater. 1977. Patterns of Welfare Use.
   Joint Center for Urban Studies, Working Paper no. 77. Cambridge, Mass.: Joint Center for Urban Studies.

Rosa, Jean-Jacques. 1982. "France." In The World Crisis in
   Social Security, edited by Jean-Jacques Rosa, Richard Hem-
   ming, John A. Kay, Oronato Castellino, Noriyuki Takayama,
   Ingemar Staht, Martin C. Janssen, Heinz H. Müller, Sherwin
   Rosen, Karl Heinz Jüttemeir, and Hans-Georg Petersen, pp.
   10-28. San Francisco: Institute for Contemporary Studies.

Ross, Rosalind Brooke, and Hans F. Zacher. 1983. Social Legislation in the Federal Republic of Germany. London: Bedford
   Square Press.

Rutter, Michael, and Nicola Madge. 1976. Cycles of Disadvantage:
   A Review of the Research. London: Heinemann.

Ryan, William. 1971. Blaming the Victim. New York: Pantheon
   Press.

Rydenfelt, Sven. 1984. "Today's Sweden Looks to the Entrepreneurs." Wall Street Journal, December 5, p. 31.

_____. 1982. "The Negative Side of Sweden's Welfare." In The Welfare State: Opposing Viewpoints, edited by David Bender, pp. 34-37. St. Paul, Minn.: Greenhaven Press.

_____. 1981. The Rise and Decline of the Swedish Welfare State. Lund: National Economics Institute, Lund University Press.

Salamon, Lester M. 1978. The Elusive Consensus. New York: Praeger.

Schaber, Gaston. 1980. Pauvreté persistant/grande région. Project no. 20. Walferdange, Luxembourg: Groupe Étude pour les Problèmes de la Pauvreté.

Schanche, Don. 1982. "Italy: Social Security." Los Angeles Times, February 28, pt. 5, p. 3.

Schmidt, Joseph. 1982. "A Portrait of Poverty in the Wealthy Welfare State, Germany." German Tribune, November 14, p. 15.

Schneiderman, Leonard. 1974. "Value Orientations and Preferences of Chronic Relief Recipients." Social Work 9:13-18.

Seewandono, Iwan. 1984. Private correspondence with the author, October 22. (Seewandono is on the Sociology faculty of Erasmus University, Rotterdam.)

Segalman, Ralph. 1976. "Theoretical Models of Social Structure and the Practice of Social Work." Arête 4:37-50.

Segalman, Ralph, and Asoke Basu. 1981. Poverty in América: The Welfare Dilemma. Westport, Conn.: Greenwood Press.

Sharff, Jagna Wojcicka. 1981. "Free Enterprise and the Ghetto Family." Psychology Today, vol. 15, no. 4 (March).

Sheehan, Susan. 1976. A Welfare Mother. New York: New American Library, Mentor.

Shkuda, Anne N. 1976. Former Welfare Families: Independence and Recurring Dependency. New York: New School for Social Research, Center for New York City Affairs.

Simanis, Joseph G. , and John R. Coleman. 1980. "Health Care Expenditures in Nine Industrialized Countries: 1960-1976." Social Security Bulletin, January, pp. 3-8.

Sklar, June, and Beth Berkov. 1974. Teenage Family Formation in Post War America. Population Reprint no. 447. Berkeley, Calif.: Institute of International Studies, International Population and Urban Research.

Smarz, Zofia, et al. 1983. "The Welfare State in Crisis." Newsweek, July 25, pp. 8-14.

Sociaal en Cultureel Planbureau. 1980. Poverty in the Netherlands: A Report to the Commission of the European Communities. Rijswijk: Sociaal en Cultureel Planbureau.

Social Security in Britain. 1977. London: British Information Service.

Stack, Carol B. 1974. All Our Kin: Strategies for Survival in a Black Community. New York: Harper & Row.

Stone, Robert C. , and Frederic T. Schlamp. 1965. "Characteristics Associated with Receipt and Non-Receipt of Financial Aid from Welfare Agencies." Welfare in Review 3: 1-11.

Stovall, Sten. 1980. "Denmark's Welfare System Proving Costly." Los Angeles Times, December 1, pt. IB, p. 2.

Strang, Heinz. 1984. Sozialhilfebedurftigkeit: Struktur—Ursachen—Wirkung unter besonderer Berucksichtung der Effetivitat der Sozialhilfe [The quality of social welfare dependency: structure, sources, and operation of special factors related to the effectiveness of social assistance: research report]. Forschungsbericht, Hanover: Institute for Social Pedagogy of the University of Hildesheim.

_____. 1970. Erscheinungs formen der social Hilfebedurfkeit: Beitrag zur Geschichte Theorie und empirischen Analyse der Armut. Stuttgart: Ferdinande Enke.

Tifflin, John. 1984. "Dutch Treat." Sixty Minutes. Columbia Broadcasting System, September 9.

Townsend, Peter. 1979. Poverty in the United Kingdom. Berkeley and Los Angeles: University of California.

Trimborn, Harry. 1982. "West Germany Reassessing: How the Social Safety Net Is Holding Up Abroad." Los Angeles Times, February 28, pt. 5, p. 3.

Tyrell, R. Emmett, Jr. 1977. The Future That Doesn't Work: Social Democracy's Failures in Britain. New York: Doubleday.

U.S., Department of Health and Human Services. 1983. Research on Long Term Dependency: Statement of Work. Request for Proposal, no. 746082. Washington, D.C.: Government Printing Office.

U.S., Department of Labor, Bureau of the Census. 1974. "Poor Families Headed by Women Show Rise." Los Angeles Times, April 5, pt. 1B, p. 3.

U.S., Office of Management and Budget. 1984. Study of Policy Options for Reducing Long-Term Welfare Dependency: Statement of Work. Department of Health and Human Services, OMB nos. 0990-0109, 0990-0155. Washington, D.C.: Office of Management and Budget.

Van Doorn, Jacques. 1978. "Welfare State and Welfare Society: The Dutch Experience." Netherlands Journal of Sociology 14: 1-8.

Walliman, Isidor. 1983. "The French Welfare System: Can Mitterrand's Decentralization Policy Solve Some of Its Own Problems?" Paper presented at meetings of the American Sociological Association, Detroit, August 21-September 4.

West, Donald J. 1984. Delinquency, Its Roots, Careers, and Prospects. London: Heinemann Educational.

Widman, Lyn. 1983. "The Welfare State in Crisis." Newsweek, July 25, pp. 8, 9, 10, 11, 14.

Wilson, William Julium. 1980. The Declining Significance of Race. Chicago: University of Chicago Press.

Wishnov, Barbara. 1973. Determinants of the Work-Welfare Choice: A Study of AFDC Women. Boston: Boston College Press.

Wright, Catherine H. 1955. "Problem Families: A Review and Some Observations." Medical Officer, December 30, pp. 94, 381-84.

2

SWITZERLAND AS A
WELFARE STATE

When people think of a welfare state, countries such as Sweden or England come to mind. With these examples comes the notion of cradle-to-grave services and security for everyone, where incomes are equalized by taxation and benefits and where a relatively comfortable life is available to everyone without too much effort on anyone's part. Thus, the idea of Switzerland as a welfare state is unusual. Especially from the viewpoint of general income equalization, Switzerland can hardly be called a welfare state. Yet, the thrust of the Swiss economy is far from economically punitive toward individuals in the society. The emphasis of Swiss taxation and government policy generally is one that encourages the accumulation of capital and the expansion of private industry. In that respect, also, Switzerland is hardly a welfare state. The socialist notion of nationalization of industry applies only to the Swiss railways and post office, and even here the Swiss are different, in that these two industries either break even or come close to it.

So, how is Switzerland a welfare state? Is it a welfare state in that it has designed and carried out a policy that seeks to prevent poverty? Is it a welfare state in that it succeeds in controlling poverty and in preventing its growth? In this respect, if one were to define welfare state as a state that seeks to eliminate poverty and the factors that cause it, one might name Switzerland as the ultimate welfare state.

In the other so-called welfare states in many of the industrialized nations, a sizable native population is subsidized by welfare or by covert unearned social insurance or other income-redistribution mechanisms. Wilensky (1975) lists four general thrusts of the welfare state. The first is that the state advocates or supports universally available social welfare through a compulsory program of public assistance for the poor, aged, unemployed, and so forth. The second is a

48

redistribution of wealth through severe taxation from rich to poor, major land reform, and equalization of all incomes on an immediate basis. The third is economic planning and advocacy of government prescription of the level and nature of resource allocation and commodity production and distribution. The fourth is government ownership of the means of production. Wilensky has also added advocacy of equality of opportunity in addition to the above four thrusts of the welfare state.

The nations that have implemented this five-pronged program are now presented with the problems of high social costs: often a lowering of quality and services, a falling gross national product, a lowering of service availability, rising inflation rates, a falling off of worker productivity, and often an exodus of capital to other less-taxed countries. Along with this has come a variety of social problems such as high rates of drug addiction, alcoholism, increased unmarried motherhood, juvenile delinquency, and high suicide rates (see Painton and Malkin 1980; Seeger 1981).

Under the welfare-state model, the onus for support of individuals and families on a relative-deprivation basis rests with the government. Under the policy, no pressure is exerted on those individuals who prefer to collect benefits rather than to prepare for, seek, and hold employment. Similarly, in the welfare state, with the onus of responsibility for care on the state, there is little or no financial constraint on the individual for support of his or her children, whether conceived in or out of wedlock, or of aged parents. Child-custody payments for children of prior marriages and responsibility for children conceived outside of marriage are not generally enforced when state subsistence is available to all children and their families.

In the welfare states, by and large, the distinction between social insurance and public assistance has become blurred, and purposefully so. With the government ready and willing to accept responsibility for a floor of income security for all, the concern that some individuals might be reluctant to apply for aid because of the "shame" or stigma attached to aid has caused the governments to provide unearned income guarantees within the social-insurance mechanism, naming it a benefit rather than an aid. The emphasis of the distributive agency is on provision of such supports, with only limited constraints beyond those relating to the specific eligibility requirements for the program.

Despite the fact that Switzerland is not found on the list of nations that consolidate their social-insurance and public-assistance programs and that place a heavy emphasis on income redistribution, one should not conclude that Switzerland is a backward nation in regard to social security. A study of the Swiss social security system in comparison with nine other major industrial nations (including the United Kingdom, the Netherlands, Austria, Sweden, the Federal Re-

public of Germany, Belgium, and the United States) indicated that
Switzerland provides the highest old-age insurance benefits after
Sweden after deduction of taxes and social-insurance contributions
(Union Bank of Switzerland 1977). An analysis of net retirement benefits among the ten nations shows 95.6 percent of earnings for Switzerland, compared with 84.3 percent for West Germany and 80.4 percent
for the Netherlands. In the case of survivors' benefits, Switzerland,
based on performance related to earnings, exceeded other national
averages by 50 percent, and the ratio between net survivors' pensions
and the lost net earnings was 133.3 percent, compared with 106.1
percent for Sweden and 100 percent for Finland. In relation to invalidity based on sickness and where the income equals the national average, the Swiss citizen receives invalidity benefits that actually exceed
the lost net earnings by 9.8 percent. This is topped among the ten
nations only by Sweden, at 34.9 percent; Finland, at 11.5 percent;
and the United Kingdom, at 14.8 percent. If the invalidity is caused
by accident, however, the picture is changed by supplementary benefits in other nations. An extreme example of this is Austria, where
the ratio of net disability benefits to the lost net earnings goes up
from 78.3 percent for sickness invalidity to 173.1 percent for accident invalidity. (This seems to place a premium on accident invalidity.) Health-insurance benefits in Switzerland were also far above
average for the nations. Health insurance provides 80-100 percent
of hospital-ward confinement, and nonhospital costs are insured by
the social-insurance programs in all of the countries except the United
States. Swiss health benefits are also comprehensive and show little
variation from the other nine nations, except in the instance of loss-
of-earnings compensation for illness, for which most Swiss workers
carry private insurance and for which employers are required to continue wages or salaries for at least one month of incapacity.

Compared with net earnings, the Swiss rate of unemployment
compensation exceeds that of most other countries in the survey. In
the medium-income category, Swiss unemployment compensation
amounts to 90 percent of the lost net earnings, exceeding that of West
Germany, Sweden, and the Netherlands and far exceeding Canadian,
U.S., and Austrian levels. In the cases of family allowances and
tax relief for large families, the allowances and relief are somewhat
less generous than most of the other countries, except for the United
States, which has never introduced a family-allowance system.

Despite the generally comprehensive and relatively liberal social insurance provided in Switzerland, the deductions from earnings
in terms of taxes and social security contributions are hardly higher
than Austria, the United States, or Canada. The total deductions (or
load on earnings) range from 14.4 percent in Switzerland to 12-15
percent in the other three countries. (Sweden has equivalent rates
of twice that level.)

The survey under discussion was conducted in 1977. The Swiss level of actual benefits will probably be much higher (in 1985) now that the "Second Pillar" of Swiss social insurance, namely, the employer-related company pensions, has become mandatory for all employees.

The Swiss social-insurance system also reflects welfare-state principles in that a standard ceiling on insurance-covered employment is the basis for retirement and survivors' benefits, just as in other nations, and in that contributions paid by employers and employees for social security apply to the total wages or salary, regardless of amount. Thus, a bank clerk and bank president, earning respectively, Fr20,000 and Fr200,000 per annum, each pay 4.7 percent (employer and employee), making a total of 9.4 percent. Thus, the bank clerk and employer pay a total of Fr1,880 and the bank president and employer pay a total of Fr18,800. It should be noted, however, that both collect the same amount of pension, namely, Fr13,200 worth of maximum benefits, plus 50 percent more if they are married. In this sense, therefore, the Swiss social-insurance program can also be viewed as an income-redistribution system.

The Swiss plan for social security is dependent upon three pillars. The First Pillar is federal social security funded primarily by employer and employee contributions. The second is the retirement pension programs established by most employers with employee participation. The Third Pillar constitutes the private life and retirement insurances of the individual and his or her savings.

For every gainfully employed person who makes Fr16,600 per year, the minimum retirement pension (for the first pillar only) amounts to Fr9,960 per year (Schweizer 1985). The provision of a social component in the system providing a minimum pension for persons whose costs do not fully meet the requirements by wage deductions also makes the Swiss system similar to other welfare states.

Wilensky's data (1975) also support the conclusion that Swiss social security expenditures hold up well in comparison with that of the United States, Iceland, Canada, Australia, and Japan. In 1966 Swiss expenditures for social security were 9.5 percent of the gross national product, compared with 7.9 percent for the United States and 6.2 percent for Japan.

In yet another sense, Switzerland may be considered even more of a welfare state than most others, in that it seeks to be compassionate to the needy not merely by providing aid, but by ensuring that all of its residents are provided with employment, prepared for a life of self-sufficient employment, and provided with a level of living that surpasses all other countries. Switzerland is able to do this by maintaining the lowest unemployment rate in the world, about 1.1 percent (Schweizer 1984). Its inflation rate is 3.0 percent (Union Bank of Switzerland 1984), and its per capita income is Fr27,000, or US$12,000 (Schweizer 1984).

Galbraith (1977) believes that the Swiss success as a practical welfare state is a result of Swiss collective responsibility and intelligence, in which the citizen rather than the leader holds the key. Swiss citizens do not delegate; instead, they take responsibility at the local level. Galbraith believes that the Swiss have three sources of strength that make the Swiss system possible. The first is that each participant in the democracy has a personal concern for the result. One person's vote and voice can have an appreciable bearing on the outcome as long as the local community and region retain autonomy and accept full responsibility and authority for most government operations. Only that which cannot be managed by the local community is relegated to the canton, and only that which cannot be carried out by the canton is delegated to the confederation (the Swiss federal government). This sense of community with the responsibility of the individual preceding the community, that of the community preceding the canton, and that of the canton preceding the confederation is what Galbraith describes as the second Swiss strength. The third strength is the Swiss interest in results rather than in principles.

It is this Swiss trait of concern for results rather than for principles that has caused the Swiss to hold back on adoption of the welfare-state principles as they relate to public welfare policy.

Both citizens and policy makers in Switzerland seen to have a different view about aid to needy persons in their country compared with the welfare-state ethos of a minimum guaranteed income for all on the basis of legalistic entitlements. The Swiss viewpoint might be described by some liberals as medieval and cruel. Others will recognize in it the policies out of which the Elizabethan Poor Law first developed.

First, it should be understood that Swiss neighborly concern for the poor is not the charity of the early church, which put a responsibility on all to give something (anything at all) to the poor without any real provision for their security in the long run. Neither is the Swiss view an advanced model of "social justice," involving a modern mixture of socialism and tax support, that seeks a restructuring of the society and an equalization of goods for all. It is also not the modern liberal view, which demands that the government aid the poor by taxation.

Whatever the reason, whether it is a heavy remnant of gemeinschaft (community) concern and responsibility or whether it derives from the Swiss business sense, the Swiss approach to problems always relates to the question, What needs to be done to change the situation so that the problem will be resolved rather than alleviated or suppressed? This, of course, also reflects the concern of Swiss leaders that if a problem is not really resolved, it will still be around to plague them on the morrow (Gruner and Junker 1978). So, the Swiss, in the manner of the highly inventive minds of industrial revolutionary times, arrived at a seemingly rational assignment for social policy. That assignment is strongly tied to utilitarianism (as was much of the economic mechanism of the industrial revolution). This policy first had to meet the requirement of aiding the poor in such a manner that they would be temporarily helped and yet not be encouraged to become dependent on that aid. Second, aid had to be tied to a policy of encouraging the poor to help themselves as much as possible and of moving them as rapidly as possible out of poverty (if not in that generation then, at least, in the next). Still another requirement was related to prevention of poverty—to encourage people to provide for their times of need (Holzer 1971). But most people, unfortunately, are afflicted with what Karen Horney (1942) called the neurotic exemption, in that in good times they never expect to fall into bad circumstances. Because of this, the Swiss installed a far-ranging and expanding program of compulsory social insurance, which places on each worker and employer a compulsory shared-risk program to provide money for most of the workers for later expected and unexpected financial needs. Thus, the primarily self-earned (with a minor government contribution) insurance program provides the worker with old-age retirement, disability and sickness insurance, survivors' insurance, unemployment compensation, and more. Unlike the social-insurance policy of many other industrialized nations, these programs are so designed that the beneficiary cannot control the outcome. Unemployment compensation is given only if the person is validly out of work and is really available for employment as it opens up. Disability is strictly defined to such a degree that it does not become an alternative to work, as has occurred in many other nations. Social insurance in Switzerland is operated as if it really were an insurance program, and the only difference between this program and commercial forms of insurance is that it covers almost everyone. No commercial insurance company would remain solvent if clients were allowed to insure themselves against a house fire and then develop a hobby of collecting inflammables. So, no social insurance program can remain viable, generally unsubsidized, and valid if the definitions and responsibilities are allowed to be amended after the fact to fit changing client demands. The lesson that Swiss social insurance has learned from the insurance business generally is that client behavior changes in relation to administrative actions.

Thus, administrative policy has to operate in such a manner as to keep people working and contributing until a real (not client-manufactured) hazard occurs. Even here, Swiss social insurance requires client rehabilitative activity (during the benefit period) toward resumption of self-sufficiency if it is possible and feasible. Otherwise, the client receives a lesser benefit (Holzer 1971; Horlick and Skolnik 1978;

Inglin, Shahel, and Tschumperlin 1979; Saxer 1965; Schulte 1980; Tschudi 1978).

The emphasis on prevention via social insurance is matched with heavy emphasis on prevention via preparation for employment. No one, seemingly, is exempt from effective schooling and occupational training and experience (see Casparis and Vaz 1979). If a child should begin to falter in preparation for adult self-sufficiency, the whole community becomes concerned—not just the school's personnel, but the whole gamut of formal and informal social-control mechanisms become involved. Truancy is apparently little known in Switzerland, and children not in school during a school day would become the concern of the local police, the neighbors, the clergy, and anyone else who might be around. The educational-occupational offerings are varied and broad. There are offerings at every level, far into adult education years. The old concept that Switzerland has few resources other than its people is matched with a communal and cantonal policy that ensures that resident children are educationally and technologically prepared to take their places in the productive economy.

Prevention of native poverty is also ensured by strict control of immigration (Walliman 1974). The Swiss have long realized the nature of the limits of their national resources. They hold to the view that they cannot provide for everyone in the world or even for everyone who would seek to come to Switzerland, so they have carefully selected a limited scope of responsibility. These limits, whatever the rest of the world may think of them, begin with a strong concern for their own. In a sense, the Swiss have sought to solve the traditional dilemma of balance as posed by Hillel the Elder of Jewish history, namely, "If I am not for myself, who will be for me?"—but, "If I am only for myself, what am I?" (See Hertz 1945.) In confronting this juxtaposition, the Swiss have weighted their decision strongly on the former, with limited concern for the latter. This may seem harsh and inhumane, but it is probably no less harsh a result than exists in other countries where policy is weighted toward the alternative and where intergenerational dependency on government aid has become a pattern among some of the population (Strang 1970). Thus, the Swiss policy on immigration is a stringent one, operated almost as if it were controlled by a calibrated spigot that is opened only when negative unemployment exists and is quickly shut when Swiss natives begin to draw excessively on unemployment compensation.*

------

*In actuality, immigration controls are automatic rather than under direct governmental action. The maximum number of temporary immigrants under law is based on a national formula. Within this formula, each canton is permitted a limit based on population for-

Despite all of this preventive effort, some Swiss do find themselves in poverty. Such persons are served not by the federal social-
insurance program, but by the local cantons and communities. The
differences between Swiss social insurance and public assistance are
carefully maintained, because these differences serve a necessary
function.

Swiss social insurance, public welfare, immigration policy,
and related factors will be presented at length in ensuing chapters.
In the remainder of this chapter, some of the basic data about Swiss
development that has helped to shape its effective economic and social
program will be presented.

Geographically, Switzerland is surrounded by West Germany,
Austria, Italy, and France, without direct access to the sea. The
land area is about 41,000 square kilometers, which makes it comparable to Belgium or Denmark. The Swiss population totals almost six
and one-half million, of whom about nine hundred thousand (14 percent)
are foreign nationals. Switzerland thus has one of the highest proportions of nonnative populations in the Western world. Population density
is about one hundred and fifty-seven persons per square kilometer
(Schweizer 1984).

Despite its four national languages (German, French, Italian,
and Romanish) and despite the multiplicity of social, political, and
religious differences, Switzerland is viewed as a cohesive unit by
most external observers. Unlike this view from the outside, Switzerland consists of 26 separate and generally independent cantons. These
cantons are subsivided into over twenty-two hundred communities or
communes, each of which retains a high degree of independence.

The original Swiss Confederation (of cantons) was organized in
1291, primarily for mutual defense. In 1848 a national government
was organized that had limited functions and powers and in which the
cantons and local governments retained considerable importance and
autonomy. According to Clark (1979), "Switzerland's local and regional governments probably have a greater share of power than in
any other nation in the world" (p. 16). Because of Switzerland's decentralized political system, the bulk of Swiss taxes are imposed and
spent by the individual cantons and communities.

Today's confederation is governed by an executive council of
seven members and a two-chamber legislature (Schweizerische Bundeskanzlei 1985). The Standrat is a representative body made up of
two delegates from each canton, and the Nationalrat is a chamber of
200 members elected directly on a proportional population basis.

---

mulas. Thus, the controls are related to population and employment
changes as they occur in each canton and in the nation.

Both chambers have equal status. The confederation is limited in power to operate only in realms conferred upon it by the cantons and the electorate. Currently, these realms include internal and external security, foreign affairs, customs, immigration, postal and communication services, monetary controls, railroads, national defense, roads and water navigation, social security, agricultural support, and other limited authorized concerns (Kummerly and Frey 1984). One of the mandates of the confederation is to uphold the constitutions of the cantons.

Each canton and most of the communities have their own constitution and laws. All Swiss citizens are equal before the law, although citizenship in the nation derives from birth or naturalization within the individual commune and canton.* The governance of Switzerland is limited by the rights of citizenry and by the initiative and referendum processes. Each canton and commune government is directed by its own legislative body elected by its citizenry. Each commune and canton allows every citizen maximum participation in the way in which his commune and canton is run (Kummerly and Frey 1984). Each commune collects taxes as well as serves the canton and confederation in terms of indirect taxation.

C. F. Ramuz is reported by Kummerly and Frey to have said of Switzerland, "It is an overwhelmingly difficult task to describe a nation: even more so when it doesn't exist" (p. 44). Political organization operates not only through 11 parties (including 4 major parties) but, more recently, has been influenced by a variety of "citizens' committees" united over particular issues or serving as pressure groups. Thus, political action is slow and requires years for legal changes to occur. Changes do take place only when a sizable consensus has been achieved.

Almost three million people (46 percent of the population) are actively engaged in economic activity as employees or employers.

---

*Citizenship can be gained as well by marriage of a woman to a male Swiss citizen or by application of a foreign parent on behalf of a child who was born and reared in Switzerland—providing this is done before that child's 18th birthday. There are actually three citizenships—commune, canton, and confederation—and each cannot exist without the others. Citizenship begins, in each case, at the communal level.

Recent legislation has removed the possibility of citizenship by marriage to a male citizen. Instead, the foreign spouse of a citizen (male or female) merely becomes eligible for immediate residency and then can apply for citizenship through regular local community channels after five years of acceptable residency.

Representing 23 percent of the working population, 690,000 foreigners are employed in the economy; 7.2 percent of the working force are involved in primary industries, such as agriculture and forest industries; 39.8 percent are involved in the secondary production sector, including handworkers, factory workers, and others; and 55 percent are involved in the service industries (Schweizer 1984).

Switzerland's economy is surprisingly vigorous for a nation that is two-thirds covered by forests and rock and has a dearth of raw materials. Schweizer reports a gross national product for 1983 of Fr179 million, equivalent to US$78,500 million. Part of Swiss economic success is derived from the Swiss pattern of industrial relations, under which employers and employees negotiate for periodic contracts without having to resolve their conflicts by strikes. Since the early World War II years, when Switzerland found itself surrounded by Nazi-held territories, management and labor have sought to arrive at a method of resolving their differences. A method was voluntarily worked out: an arbitration system with severe penalties instead of government involvement. The voluntary system has worked so well that it has been continued over the postwar period and into the 1980 decade (Burkhardt 1978).*

The postwar years have resulted in experience for Switzerland that is decidedly different from the experience of other Western industrialized nations. Clark (1979) indicates that Switzerland is the only nation that has succeeded in industrializing without having to create huge cities that bring all of the social problems allied with urban development. The Swiss economy is not burdened with a sizable unemployed (and unemployable) population, and the Swiss inflation rate is generally lower than most other developed nations. Clinard (1978) reports that Swiss communities have almost no violent crimes and are relatively crime free when compared with other industrialized nations. The Swiss public-assistance systems (which we will later describe) are humane and compassionate and relatively free of the welfare dependency prevalent in other welfare states. The Swiss economy, un-

---

*The story of Swiss industrial peace and consensus is reported in detail by Jürg Steiner (1974). Consensus is sought at all levels of society. In the political realm no action is taken until a general consensus has been achieved. The concern for action by consensus is exemplified by policies of the Federal Council. Decisions must be arrived at by the entire group rather than by each individual councillor. Councillors must abide by the consensus and remain in office even if they fall into disagreement with the other members of the council. Not to remain and carry out the group decision is considered "bad faith."

like the economies of other Western nations, is neither overburdened
by excessive national defense expenditures nor the victim of military-
industrial machinery that serves to distort the political process.
Also unlike many other Western nations, the Swiss economy is bol-
stered by deposits and investments from people who trust the Swiss
banks more than they do their own. Switzerland is not beset with sizable proportions of unemployed youth with inadequate education,
training, and work conditioning; Switzerland has no underground economy of unreported employment, sales, and untaxed income. The net-
earnings level (Union Bank of Switzerland 1982) shows that net earnings
in Zurich and Geneva match those of Chicago and San Francisco and
surpass those of New York, Los Angeles, Montreal, and all other
metropolitan centers. Economic reports indicate a very high level of
personal savings and capital investment. Siegenthaler (1978) reports
a trend toward increased wages and salaries for groups that were disadvantaged in previous decades. Unlike the mass of workers in many
other industrialized nations, the workers of Switzerland are still imbued with a belief in the Protestant ethic, which values hard work,
sobriety, avoidance of ostentation, honesty, and conscientiousness
(Kramer 1980). In 1976 the population of Switzerland, including its
employee groups, resoundingly rejected an initiative to cut the work-
week from 44 to 40 hours. In terms of income taxes, Switzerland's
levy on workers is among the lowest of the ten nations studied by the
Union Bank of Switzerland (see Table 2.1).

So, why is Switzerland so different? Why does it present exceptions along all of the parameters noted above?

It may be claimed that the Swiss experience is unique and has
no relevance to social-control problems in other industrialized countries. The reasons usually given for Swiss immunity to social ills
include its geographic isolation, its small population, the strong influence of its religion, and its homogeneity. In the reality of the closing
twentieth century, however, Switzerland is not that much different
from other European states. Modern transportation and communication have made the Swiss no more isolated than any other European
nation; their population of almost sixty-five million is greater than
that of Denmark, Norway, and Ireland and is comparable to that of
many other European nations (its population is larger than 55 percent
of all of the countries in the world); other countries in Europe have
strong religious ties, such as the Netherlands, Belgium, and Sweden,
and yet suffer much of the modern malaise of social disorganization.
Most Swiss would challenge the claim that their population is homogeneous, citing differences in language, culture, ethnicity, regionalism,
and localism, as well as religion. Luck (1978) reports that the things
that potentially divide the Swiss are very real.

If it is not small size or religion or geographical isolation, what
then makes Switzerland immune to the modern industrial Pandora's

TABLE 2.1

Income Tax as a Percentage of Gross Earnings across Countries

Country	Income Category*		
	1	2	3
Belgium	8.4	13.9	22.6
West Germany	9.8	12.6	19.0
Finland	15.4	24.2	33.8
Great Britain	16.5	18.4	21.7
Canada	8.3	15.9	24.3
Netherlands	9.3	13.4	25.2
Austria	0.1	6.4	15.3
Sweden	29.7	37.9	49.4
Switzerland	3.1	6.7	13.4
United States	6.7	12.4	19.8

*Income category 1 equals the annual average before any deductions in wage or salary of Fr24,200 ($9,798 as of 1977). Income category 2 is equal to 150 percent of category 1, representing annual wage or salary of Fr36,300 ($14,696 as of 1977). Income category 3 is equivalent to 250 percent of category 1, representing an annual wage or salary of Fr60,500 ($24,494 as of 1977).

Note: Wilensky indicates that the total tax "take" for New York and California for 1965–66 was 36 percent and 34 percent, respectively, well above the average "take" for the United States of 28 percent and above that of the United Kingdom and Denmark. Thus, the Swiss tax pressure is comparably lower than for most nations at all levels, except for Austria.

Source: Union Bank of Switzerland, Social Security in Ten Nations (Zurich: Economic Research Department, Union Bank of Switzerland, 1977), p. 26.

box of social problems? Some of the answers will be presented in the balance of this chapter; others will become evident later.

One of the keys to the Swiss solution seems to be based upon retention of political power in the local and cantonal units rather than in the federal structure. Apparently, the Swiss confederation depends on the countervailing of fierce local loyalties and a rational need for national cooperation, providing a set of mutually reinforcing tensions that tend to cancel each other out (Frenkel 1978) by producing a network of carefully designed compromises. The Swiss cohesion, according to Frenkel, is in large part dependent upon a historical, rational, mutual concern for protection against invasive neighbors (Schmid 1981). In modern terms that translates into a need to maintain economic viability. Historically, the Swiss made careful choices in the face of adversaries. For many years, during the agricultural period, there was an export of excess population in order to keep family farms from being subdivided into uneconomic units. For the same reason, many of the younger sons left Switzerland to hire out as foreign mercenaries, making possible the subvention of the home farm from their earnings. (The papal Swiss guard is a historical remnant of this practice.) Many Swiss sons traveled overseas as commercial representatives of Swiss products. The national defense of Switzerland has been traditionally comprised of reserve citizen soldiers rather than a large standing army with expensive equipment. Another centralizing force has been the general acceptance of an ethos of hard work, free enterprise, honest relations, insistence on self-support, and the building of a pattern of dependability without which the Swiss banks and businesses would not be possible. It is this reputation for dependability that has made the Swiss banks the model of an agency of trust. Swiss government, particularly at the local and cantonal levels, is astonishingly depoliticized "because important local issues should not be controlled by political ploys" (Frenkel 1978, p. 123). According to Schmid, "The Swiss have tended to see politics as simply a process for working out commonsense solutions" (1981, p. 136) to the problems that face them. By and large, all major decisions are made in the cantons, and the elected representatives are held responsible by the citizens of the cantons. Although the practice is shrinking, a few cantons still hold open-air meetings where the representatives are called to account by their neighbors, both for decisions made on the local level and in the national parliament. Because the citizen is tied first to his commune, then the canton, and only then to the nation, this leads to more responsible behavior, even for persons in the political realm. The position taken by the individual usually has direct personal consequences, whether in the form of increased or decreased taxes or in terms of individual authority and responsibility. The tendency to support an ideological position or a stance that shifts the costs to the nation generally is seldom encountered (Gretler and Mandl 1973). When the nation is called on to intervene in a program, it is only because the cantons view the task as unmanageable at a local level. The major assignment of establishing and maintaining a social-insurance program was seen as a task for the nation, but the care of individuals with problems of self-sufficiency continues to be a local responsibility. Thus, Swiss social policy is a product of the joint efforts of a confederation of equally powerful cantons rather than a program installed from above by a national entity (Gretler and Mandl 1973; Sorrell 1972).

Government officials, according to Frenkel 1978), operate in a nonindividualized manner, seeking to solve problems and to resolve issues rather than to make any one individual look "good." Even the presidency of Switzerland rotates year to year among the members of the Federal Council, in order to maintain a government of law and consensus rather than one based on individuals. Thus, Switzerland has chosen not "to put her faith in princes." Luck (1978, p. xii) reports that the Swiss are "so fiercely cantonal in their loyalties that they vigorously resist the encroachments of the federal government. [Yet, when they] meet the same people in a foreign country . . . they are Swiss and proud of it." The Swiss defend not only the governmental autonomy of their communal and cantonal government, they also defend the social institutions of their locality as well. In some instances, federal subventions were made to mutual aid or philanthropic associations rather than to individuals, primarily because grants to individuals might weaken local cohesion (Schar 1978).

Because of the accepted dependability of government and banking in Switzerland, Swiss residents have one of the highest rates of personal savings among all the nations, as well as a high rate of savings and protection by private insurance. This high rate of savings, deriving from lower taxes and high levels of confidence in Swiss institutions, provides added capital for commercial and industrial development. Such confidence in government, of course, would not be possible if the government were not locally at hand and subject to control by local citizenry. Local budget decisions in most places require a vote of approval from the electorate before taking effect.

Switzerland has also avoided many of the social problems that in other countries emanate from neglected, uneconomic city slum areas. Because it has avoided urbanization while enjoying the benefits of industrialization, it has no slum areas with crime, welfare dependency, drug markets, educational evasion, fire hazards, and other social ills.

A number of social patterns also strengthen Swiss cohesion and consensus. Much of the local consensus over issues is strengthened by the frequent contact of local people in their military drills and exercises, in the local civil defense activities, in the operations of the volunteer fire departments (which exist in most towns other than in the few large cities), and in the variety of social clubs and guilds. (It has been estimated that almost every Swiss is a member of at least one such guild, and many have multiple memberships of three, four, and five.) Civic organizations, such as Rotary, take on much more importance in Europe than they do in the United States, and many Swiss residents come together in these groups to discuss local, cantonal, and national issues. The general lack of social-class distinctions among the Swiss leads to more frequent communication among local residents. Despite the large numbers of aliens in the society, many of them are in the category of legal resident, having been in the country for more than ten years, and this leads to their involvement in local issues as well. Finally, the local retention of such services as police, schools, family-relations courts, and civil courts provides a strong influence by the resident population on agencies of social control. Thus, a high level of consensus and conformity to community norms is maintained, along with frequent opportunities for the expression of differing viewpoints.

There is strong support of family responsibility and authority in the Swiss communities. There is also strong support of the principle of subsidiarity, which retains responsibility and authority for family, for commune, and for cantons.* This principle requires that no function shall be transferred, unless the family or commune or canton cannot successfully carry out that function at the local level. This puts a priority on the family and community and provides the community with a "first call" tax base for carrying out its functions (Lüscher, Ritter, and Gross 1973).

Luck (1978) reports that the Swiss approach to self-governance and social policy is truly a unique one and has much to offer other nations in the resolution of their problems. It is especially so in demonstrating a model of a nation that provides the promised benefits of the welfare state without offsetting debilitating costs and antiproductive incentives.

REFERENCES

Büchi, O., A. Inglin, A. W. Shahel, and P. Tschümperlin. 1979. "Die Organisation des Sozialwesens in der Schweiz" [The organization of social policy in Switzerland]. Report for the Tripartite Study Conference, November 19-23, Frankfurt.

Burckhardt, Lukas F. 1978. "Industry-Labor Relations: Industrial Peace." In Modern Switzerland, edited by J. Murray Luck, Hugo Aebi, Joseph von Ah, Lukas Burckhardt, Erich Gruner, and Hans Haug, pp. xi-xvi. Palo Alto, Calif.: Society for the Promotion of Science and Scholarship.

Casparis, John, and Edmund W. Vaz. 1979. Swiss Family, Society, and Youth Culture. Leiden: E. J. Brill.

*This principle is a part of the Swiss constitution.

Clark, Gardner. 1979. "Modernization without Urbanization: Switzerland as a Model of Job Development outside Large Urban Areas." Schweizerische Zeitschrift für Soziologie 6: 1-41.

Clinard, Marshall B. 1978. Cities with Little Crime. Cambridge: At the University Press.

Frenkel, Max. 1978. "Swiss Federalism in the Twentieth Century." In Modern Switzerland, edited by J. Murray Luck, Hugo Aebi, Joseph von Ah, Lukas Burckhardt, Erich Gruner, and Hans Haug, pp. 323-38. Palo Alto, Calif.: Society for the Promotion of Science and Scholarship.

Galbraith, John Kenneth. 1977. The Age of Uncertainty. Boston: Houghton Mifflin.

Gretler, Armin, and Pierre-Emerick Mandl. 1973. Values, Trends, and Alternatives in Swiss Society. New York: Praeger.

Gruner, Erich, and Beat Junker. 1978. Burger, Staat, und Politik in der Schweiz. Basel: Lehrmittel.

Hertz, Joseph H., ed. 1945. Sayings of the Fathers. New York: Behrman House, p. 25.

Holzer, Max. 1971. "Die Sozialpolitik." In Die Schweiz seit 1945, edited by Eric Gruner, pp. 116-36. Bern: Franke.

Horlick, Max, and Alfred M. Skolnik. 1978. Mandating Private Pensions: A Study of European Experience. Social Security Administration, U.S. Department of Health, Education and Welfare. Washington, D.C.: Government Printing Office, pp. 96-114.

Horney, Karen. 1942. Self-analysis. New York: W. W. Norton, p. 115.

Kramer, Jane. 1980. "A Reporter in Europe: Zurich." New Yorker, December 15, 1980, pp. 118-35.

Kümmerly & Frey. 1984. Switzerland 1984. People and State Economy Culture. Bern: Kümmerly & Frey.

Luck, J. Murray. 1978. "Introduction." In Modern Switzerland, edited by J. Murray Luck, Hugo Aebi, Joseph von Ah, Lukas

Burckhardt, Erich Gruner, and Hans Haug, pp. xi-xvi. Palo Alto, Calif.: Society for the Promotion of Science and Scholarship.

Lüscher, Kurt K., Verena Ritter, and Peter Gross. 1973. Early Child Care in Switzerland. London: Gordon & Breach.

Painton, Frederick, and Lawrence Malkin. 1980. "Woes of the Welfare State: A Benevolent Vision Grown Too Ambitious Shrinks under an Economic Squeeze." Time (Europe edition), December 1, pp. 10-15.

Saxer, Arnold. 1965. Social Security in Switzerland. Bern: Paul Haupt.

Schar, Meinrad. 1978. "Public Health in Switzerland." In Modern Switzerland, edited by J. Murray Luck, Hugo Aebi, Joseph von Ah, Lukas Burckhardt, Erich Gruner, and Hans Haug, pp. 213-26. Palo Alto, Calif.: Society for the Promotion of Science and Scholarship.

Schmid, Carol L. 1981. Conflict and Consensus in Switzerland. Berkeley and Los Angeles: University of California Press.

Schulte, B. 1980. "Reforms of Social Security in Europe in the Periods 1965-1975 and 1975-1980." Colloquy in Perugia. Munich: Max Planck Institute for Foreign and International Justice, pp. 6-35.

Schweizer, Willy. 1985. Private correspondence with the author, January 11. (Professor Schweizer is on the faculty of the University of Bern.)

_____. 1984. "The Social Security System of Switzerland." Mimeographed. Bern: University of Bern.

Schweizerische Bundeskanzlei. 1985. Der Bund, kurz erklaert [The confederation, briefly described]. Bern: Schweizerische Bundeskanzlei.

Seeger, Murray. 1981. "Social Welfare Plans in West Europe Pinched: Countries Borrowing Heavily to Make Up for Budget Deficits, Reluctant to Increase Taxes." Los Angeles Times, February 16, pp. 1, 18-20.

Siegenthaler, Hansjorg. 1978. "Switzerland in the Twentieth Century: The Economy." In Modern Switzerland, edited by J. Murray

Luck, Hugo Aebi, Joseph von Ah, Lukas Burckhardt, Erich
Gruner, and Hans Haug. Palo Alto, Calif.: Society for the Promotion of Science and Scholarship.

Sorrell, Walter. 1972. The Swiss. New York: Bobbs-Merrill.

Steiner, Jürg. 1974. Amicable Agreement versus Majority Rule:
Conflict Resolution in Switzerland. Rev. ed. Chapel Hill, N.C.:
University of North Carolina Press.

Strang, Heinz. 1970. Erscheinungs formen der social Hilfebedurfkeit: Beitrag zur Geschichte Theorie und empirischen Analyse
der Armut. Stuttgart: Ferdinande Enke.

Tschudi, Hans Peter. 1978. "Social Security." In Modern Switzer-
land, edited by J. Murray Luck, Hugo Aebi, Joseph von Ah,
Lukas Burckhardt, Erich Gruner, and Hans Haug, pp. 199-212.
Palo Alto, Calif.: Society for the Promotion of Science and
Scholarship.

Union Bank of Switzerland. 1984. Switzerland in Figures (1984).
Zurich: Union Bank of Switzerland.

_____. 1982. Prices and Earnings around the Globe, 1982. Zurich:
Economic Research Department, Union Bank of Switzerland.

_____. 1977. Social Security in Ten Industrial Nations: A Comparison
of Social Security Costs and Benefits between Switzerland, Belgium, the Federal Republic of Germany, Finland, Great Britain,
Canada, the Netherlands, Austria, Sweden, and the United States
of America. Zurich: Economic Research Department, Union
Bank of Switzerland.

Walliman, Isidor. 1974. "Toward a Theoretical Understanding of
Ethnic Antagonism: The Case of the Foreign Workers in Switzerland." Zeitschrift für Sociologie 3, no. 1 (February):84-94.

Wilensky, Harold L. 1975. The Welfare State and Equality: Structured and Ideological Roots of Public Expenditures. Berkeley
and Los Angeles: University of California Press.

3

SWISS SOCIAL
INSURANCE AS A
CONSTRAINT ON
WELFARE DEPENDENCY

The Swiss social-insurance system is, as has been previously noted, completely separate from the public-assistance systems. The former is operated according to nationally established regulations that apply without exception or discretion under the aegis of the Federal Social Insurance Bureau. The latter are operated by the local communities or cantons with differing policies and services based on the regional needs and rehabilitative requirements of the individual clients.

The operational policies of the Swiss social-insurance system are, of course, quite different from those of the public-welfare systems in the local communities. We present the two sets of guidelines in Table 3.1.

The compulsory social insurances in Switzerland are

Old-age and survivors' insurance (AHV)
Disability insurance (IV)
Income replacement for military and civil defense conscripts
Unemployment insurance for employees
Insurance for military personnel against sickness and accident
Accident insurance for all employees
Family allowances

Also recently added has been the Second Pillar insurance for increased old-age pensions and survivors' and invalids' insurance.

The voluntary social-insurance programs include health and maternity insurance, professional superannuation insurance, and accident insurance outside the place of employment (Büchi et al. 1979).

Schweizer (1984) thinks that the belief system underlying the Swiss social-insurance system is a consensus of Swiss values, including

TABLE 3.1

Comparison of Swiss Social Insurance with Public Assistance, by Item and Function

Item	Social Insurance	Public Assistance
A. Acceptance for benefits or grants[a]	By prior inclusion on a broad-based involuntary coverage by worker and employer, with adequate period of inclusion, based on completed application form	By a carefully administered means test, conducted by a social case worker who combines examination of assets and income with "social understanding"
B. Administration[b]	Primarily by the Federal Social Insurance Bureau, on an impersonal mechanistic basis	By the local cantons or communities on a personal-contact basis, by persons able to know or to ascertain eligibility and rehabilitative case facts
C. Relative amount of benefits or grants[c]	Generally much larger than the public-aid level and based on calculated benefits earned during period of coverage, but seldom at a level equal to full-time employment (so that return to the job is encouraged wherever this is possible)	Based upon a carefully constructed individual list of subsistence needs, as related to temporary aid (except in the case of needy aged, infirm, and disabled), and as related to a carefully administered and supervised rehabilitation plan managed by community or cantonal welfare offices
D. Deduction of outside income from benefits or grants[d]	Generally there is no deduction from benefits, except where such income is related to the risk covered by the insurance (example, wage income as it relates to unemployment compensation)[e]	There is usually a careful calculation of income that is charged against the aid grant so that the actual aid grant will match the family budget, as noted under C

(continued)

67

TABLE 3.1 (continued)

Item	Social Insurance	Public Assistance
E. Stigma associated with public assistance[f]	There is no stigma attached to social-insurance benefits; the client is merely collecting the benefits he or she has earned	There is a genteel, but present, stigma attached to acceptance of public aid, especially in the case of clients who could be self-sufficient and are not aged, infirm, or disabled
F. Responsibility of relatives[g]	No responsibility of relatives is involved; the person's eligibility for benefits is based on his or her own achievements and earnings	Parents and grandparents are legally responsible for children; grown children are responsible for needy parents and grandparents; fathers and mothers of children (regardless of marriages contracted or nonmarriage liaisons) are legally responsible for their children's support; even grown siblings are held responsible when financially able
G. Funding source[h]	Primarily by the employer and employee contributions, with limited support by the government	Funded by communal and cantonal taxes
H. The right to benefits or grants[i]	A legal right exists for the benefits a person has earned	Aid grants are not considered an entitlement and are dependent upon the client's use of the grants toward rehabilitative purposes; appeal is only available to local officials and not to the courts
I. Reimbursement of benefits or grants[k]	All benefits are nonreimbursible, as long as they are based on a valid claim and on the prior required prepayments by employees and employers	Aid grants are a loan that must be repaid when and if the client is able to do so

68

aThe difference between the two approaches makes it possible to provide benefits to those who have earned them without benefits being devalued by public-aid grants. The means test ensures that public assistance will be granted only where client need is present and where it will not serve as a disincentive for client self-sufficiency.

bThe difference between these two different approaches serves to prevent confusion between earned social benefits and unearned public aid.

cThe difference between the level of benefits serves to encourage the client's return to remunerative employment where possible and as soon as possible.

dDeductions from aid grants serve to ensure that the purpose of the client family budget will be upheld and to ensure that there would be no incentive among the populace to secure assistance while earnings are coming in.

eWhen the beneficiary is eligible for more than one benefit (such as two federal insurances—old-age retirement [AHV] and invalidity [IV]—or a federal insurance and a cantonal insurance), then the benefits are adjusted to meet the defined limits.

fStigma for public assistance serves the purpose of encouraging employment and self-sufficiency where possible.

gThe responsibility of relatives is retained under public assistance to encourage familial self-sufficiency and to prevent shifting of responsibility for the needy to government support.

hFunding of financial assistance by local taxes results in involvement of the local citzienry in study and decision making on matters of public assistance. Funding of social insurance by deduction from earnings provides the recipient with a justfiable sense of having earned the benefits.

iThe reason for not making public assistance an entitlement is based on the need to have the client view the public-aid grant as part of a contract with the community, which requires participation in a plan for rehabilitation. The legal right to social-insurance benefits is also part of the fulfillment of a contract between the employee, the employer, and the state.

jThe only exception to this rule is in the case of invalidism at birth, which is an entitlement under the Ergänzungsleistungen ("supplementary security income provisions").

kThere are few instances where public-aid recipients have been pressed for repayment. The purpose of this requirement is to encourage careful client use of public-aid money. This serves to induce him or her to limit requirements.

Sources: Büchi et al. 1979; Gruner 1982; Hauser et al. 1980; Lüscher, Ritter, and Gross 1983; Meyer 1974; Schulte 1980; Social Services of the City of Bern 1981; Swiss Conference for Public Welfare 1980; Switzerland, Federal Office of Social Security 1982; Tuggener 1981; Urner 1980; U.S., Department of Health, Education and Welfare 1981; Welfare Agency of the City of Bern 1981; Welfare Agency of the City of Zurich 1978, 1979.

A responsibility of the individual to society at large,

Political participation in the solution of problems larger than those of individual concern,

A raising of individual motivation,

Strengthening of the relationship between employers and employees,

Raising of individual economic mobility.

Schweizer views the place of the individual in a society within the context that no "innocent party" should be subject to poverty and that judicial and social justice shall prevail in a society where everyone practices social responsibility.* Unlike the plans of other welfare states, however, Schweizer notes that the Swiss social-insurance system is concerned with positive outcomes. Only through an effective social-insurance system, where people receive their benefits in a just and rational manner and based on the original intent of the system, can most of the population have confidence in the institutions and depend upon them. He indicates that when such a system becomes undependable, in that unintended benefits are disbursed or intended benefits are not disbursed, then centrifugal forces can destroy the system from within. Lack of confidence in the permanence of the system can occur, as is evident in the U.S. system, where social security expenditures are mixed in with welfare grants in public reports and where social security contributory payments are dealt with by government as just another tax. This is further aggravated in the United States and in other nations when economists equate welfare grants and social security (earned benefits), as if both were part of a governmentally provided income transfer. Under such circumstances, potential beneficiaries can become concerned that they will somehow be denied their earned benefits. When this happens, one can expect the development of unreported employment and an underground economy. This does not occur in Switzerland, and the mixing of public aid and social security is not possible because they are managed by two different sectors of government.† Interviews with Swiss

---

*"Innocent party" is used here to indicate a person who has not caused his own condition of need.

†The only exceptions to this pattern are the Ergänzungsleistungen provisions for support of needy aged, infirm, or handicapped persons whose adjusted net income (after insurance fees, excess rentals, and the like) is less than Fr12,000 per annum. In such instances the person applies to the communal welfare bureau (Fürsorge), which examines the application and then processes it to the Federal Social Insurance Bureau. The periodic support check is then issued from

employees and employers show that they view social-insurance contributions and benefits with the same sense of security and confidence
as they view their personal funds on deposit in Swiss banks. Only in
Switzerland have the programs been kept entirely separate and the
distinctions kept in place.

It should also be noted that the design of each program is related to desired effects, and the blunting of effects or the misapplica-
tion of an effect upon an inappropriate condition may result in unintended counterproductive client reactions.

Swiss social insurance developed later than similar programs
in other nations. The delay can probably be considered to have been
caused by Swiss conservation in moving into new programs, as well
as by the slowness of the Swiss to allow the confederation to act on
social insurance.* Authority to legislate on sickness and accident
insurance was granted to the confederation in 1890, but it was not
until 1925 that constitutional authority was granted for old-age, survivors', and invalidism insurance. Family-allowance authority was
granted in 1945, and two years later, authority to provide for unemployment insurance and compensation for loss of earnings owing to
military service. A total of 60 years expired before the legislative
authority for social insurance was granted and even more years for
the legislation for these programs to take effect. Insurance for invalidism finally came into force in 1960.

THE OLD-AGE AND SURVIVORS' PROGRAM:
THE FIRST PILLAR

Primary among the social-insurance programs are old-age pensions. Their purpose is to provide a substantive, although minimal,
existence for retirees and for widows and orphans. The plan covers
the entire population, including employees, the self-employed, and

---

federal funds. (The Ergänzungsleistungen program serves the same
purposes as the Supplementary Security Income program serves in
the United States.)

*The history of Swiss social-insurance legislation is lengthy. In
1925 the legislature was authorized to vote on social security. A
social security law was proposed and voted down in 1931, both in the
legislature and on referendum. From 1930 through World War II,
some cantons passed their own social security programs. However,
in 1947 a new social security system was proposed and accepted on
the national level. Thus, in some Swiss cantons, both cantonal and
national social security is provided.

employers. The regulations are federal and universal for the nation. The program is carried out on a decentralized basis through 104 equalization funds of the cantons, the trade associations, and the federal government and legal agencies. As is the case with other basic retirement and survivors' funds, there is an assumed solidarity between the generations that guarantees continuity. Consequently, this program suffers from the same handicaps as similar programs abroad in that demographic changes can (and do) cause "bulges" in beneficiaries at times in years when the working populations may be smaller. But unlike its cousins in the other nations, the Swiss social security system seeks to plan for these emergent demographic trends by its fiscal and funding policies. Thus, the problem is somewhat less critical in Switzerland.

The Swiss system is a type that has been classified under the Beveridge model, similar in design to the United Kingdom, Canada, the United States, Finland, the Netherlands, and Sweden (Union Bank of Switzerland 1977). Under this model, old-age and survivors' insurance has been declared compulsory for the entire population: "All residents for these nations are . . . guaranteed a minimum, or basic pension for subsistence which is independent of the duration of contributions or previous wage and salary levels" (p. 8). Under the alternate model (the Bismarck plan, which operates in West Germany, Austria, Belgium, and France), retirement benefits are organized by occupations or professions, and the size of the pension is largely proportionate to the size of earnings before retirement. Under this model, some elements of the population are either not covered or have much smaller pensions than others. The Bismarck model also has a problem of weaker pensions in occupations that have greater risks, and these are not offset by pooling them with other, safer occupations. Similarly, the higher-paid occupations provide more generous pensions than others. The Swiss program promotes solidarity not only between generations but also among income groups by placing no ceilings on earning-related contributions and by providing that benefits at the highest shall be twice those at the lowest levels. There is also solidarity between the sexes in that there is no contribution required from nonearning married women. It also provides for equal benefits despite the greater life expectancy of women. Solidarity between single and married persons is promoted in that pensions for couples are 50 percent higher than pensions for single persons, even if the wife has paid no contribution. Solidarity between regions is promoted in that the very different incomes (based on regional differences) pay only proportionate total contributions. Also, the solidarity principle is reinforced in that town and country differences in patterns of earnings and cost of living are accounted for in the program (Charles 1983).

TABLE 3.2

Annual Full Pension Plan
(in Swiss francs)

Plan	Rate (in percent)	Minimum per Month	Maximum per Month
A. Single old-age pension	100	690	1,380
B. Married couples' pension	150	1,035	2,070
C. Widows' pension	80	552	1,104
D. Orphans' pension (one parent only)	40	276	552
E. Orphans' pension (no parents)	60	414	828

Note: The above pensions show a raise of 11.3 percent to allow for increases in prices and wages since January 1, 1984.

Source: Jean-François Charles, Social Security in Switzerland: Main Features of the Schemes and Current Problems (Report on Swiss social insurance made to the Twenty-first General Assembly of the International Social Security Association, Geneva, October 3-13, 1983), p. 181 (Bern: Federal Social Insurance Bureau).

Funding of the program is dependent primarily upon contributions from employees and employers, with some 20 percent subsidies from the confederation and the cantons.

Persons covered include all persons who have a legal residence in Switzerland as well as Swiss citizens who work abroad for a Swiss-based employer. This is compulsory insurance. In addition, Swiss nationals living abroad may voluntarily join (or retain status) in the program via consular and diplomatic representatives. All such persons are liable to pay contributions on taking up employment, but in no case later than January 1 of the year following the person's twentieth birthday. Contributions cease at age 62 for women and age 65 for men or later, as long as gainful employment or activity is continued. Employees and employers each make contributions of 4.8 percent of wages or salaries by means of pay deductions. Self-employed persons pay 9.6 percent, but if their annual income is less than Fr30,000, the rate of contribution is less, on a degressive scale to 3.46 percent at the lowest level. Persons who are not gainfully employed but who have capital assets make contributions ranging from Fr250 to Fr10,000 per annum.

Insured persons who have paid contributions for at least one full year are entitled to pensions. The amount of the pension depends

on the average annual earnings on which contributions were paid and
the duration of the contributory years. Currently, all insured persons who have paid their contributions since the program came into
force on January 1, 1948, are entitled to full pensions. The annual
full pension plan is described in Table 3.2.

In addition to the amounts in Table 3.2, monthly allowances of
Fr496 are granted for severely disabled persons requiring the aid of
an attendant. A mixed index is used to adapt the pensions to rises in
prices and wages (50 percent of the price index rise and 50 percent of
the wage index rise), adjusted every two years.

Problems of the old-age and survivors' fund derive from demographic changes, the possible introduction of a flexible retirement
age, and the possible development of pension splitting between married couples.

As of mid-1983, the Swiss old-age and survivors' fund showed
a surplus of Fr500 million. From the above data, it is easily concluded that the old-age and survivors' program in Switzerland is both
robust and effective in providing basic security for its employed
population and their beneficiaries.

THE COMPULSORY OCCUPATIONAL
RETIREMENT SCHEME: THE SECOND PILLAR

As of January 1, 1985, this program (formerly voluntary) became compulsory for all employees and employers. This program,
sometimes called the professional pensions system, is based upon
capital accumulation. Its purpose is to fill the gap between the limited
subsistence of the First Pillar and the standard of living before retirement (but at a somewhat reduced level). The First Pillar plus the
Second Pillar combined are designed to provide about 80 percent of
the income of a couple whose income prior to retirement was Fr24,000
(US$10,582) and about 56 percent of the income of a couple whose
prior income was Fr60,000 (US$26,431) (Schweizer 1984). The Sec-
ond Pillar is financed only by employer and employee contributions.
Pensions are calculated actuarily on the basis of accumulated capital.

The Third Pillar is based on personal savings, life insurance,
investments, property ownership, and so forth. The localities, the
cantons, and the confederation seek to encourage this development in
the population, especially in the case of self-employed persons for
whom the Second Pillar is often neglected. Janssen and Müller (1981)
indicate that the Third Pillar could be expected to shrink somewhat
as the First and Second Pillars become fully accepted by the covered
population, and they view this as a potential, but limited, problem.
It should be noted that as of 1983, estimates indicated that over 55

percent of the employed population was already covered under the voluntary Second Pillar plan (Charles 1983). According to Horlick and Skolnik (1978), this represented coverage of about 80 percent of all the wage and salaried workers. Only about 300,000 wage and salaried workers were not covered. Charles found that 18,100 employment institutions covering 1.6 million employees were already covered at that time. Under the new compulsory law, insurance for all employees whose earnings are over Fr14,880 and up to Fr44,640 is to be maintained. This is to be designated the "coordinated salary" and constitutes the legal minimum standard, although the institutions can (and many do) offer better provision for their members if they wish. Retirement pensions (Second Pillar) are to be actuarily calculated on the basis of total old-age capital. Contributions are to be paid in on a graduated scale based on employee's age, as follows: 25-31 years pays 7 percent; 32-41 years, 10 percent; 42-51 years, 15 percent; and 55-65 years, 18 percent. These contributions are to be income tax deductible. Unlike the practice under the voluntary plan, the new program is completely transferable among all employers. Changes in benefits based on price and wage rises are left entirely to each of the institutional funds. A national guarantee fund will be established to subsidize those institutions where the age structure is unfavorable or to protect against institutional insolvency. Periodic revisions are planned in order to bring the occupational funds to the point where the elderly, survivors, and the disabled will be able to adequately maintain their living standard.

According to Horlick and Skolnick, annual receipts from private plans were Fr6.3 billion in 1976, with reserves of 50 billion. This compares with receipts of Fr7.1 billion in contributions for the old-age, survivors', and disability programs on behalf of the 3.2 million insured in 1976. About 50 percent of the Second Pillar plans cover all three contingencies of old age, disability, and survivorship. In 1958 employers participating in Second Pillar programs were required to set up separate legal entities to protect them against loss of funds if the company was bankrupted. By 1972 vesting was made a requirement. An employee was given the right to his or her own contribution, and after five years, the right to the eventual accrued pension from both personal and employer contributions. After 30 years of service, the worker is granted the right to a full benefit. With the shift to compulsory coverage in 1985, the benefits and coverage are transferable among employers and all three risks will be covered (old age, invalidity, and survivorship).

The Swiss Second Pillar plan, now that it is compulsory, is far in advance of the U.S. system. In the United States, only a small proportion of employed workers are covered beyond social security (old-age, survivors', and invalidity insurance). In addition, the cover-

age among employers is not transferable, insurance standards are far
from uniform, and many of the funds are at risk of company actions
or misfortunes. Many such funds have been misappropriated, and
employees have often been terminated just short of becoming eligible
for pension benefits. Few of the funds have been vested. Based on
comparison with U.S. parallel arrangements, the Swiss Second Pillar
program is superior.

INVALIDITY INSURANCE

Disability payments in relation to military accidents have been
available since 1901 in Switzerland, and in relation to industrial accidents since 1912, but only after January 1, 1980, was coverage expanded to disease and congenital infirmities. Federal, cantonal, and
private organizations share in its implementation. On matters of
finance and administration, Swiss invalidity insurance (IV) rests
heavily on the old-age and survivors' program, in that the same equalization funds and their offices administer both programs. The prime
purpose of the invalidity-insurance program is the vocational rehabilitation of disabled and sick persons, including their integration (or
reintegration) into productive economic life, into social life, and
toward maximal self-care and self-sufficiency. Only if rehabilitation
is not possible is a permanent pension viewed as a continuing solution.
Rehabilitation services include medical treatment, although
curative treatment is usually the responsibility of the sickness and
accident insurance schemes, which are described later in this chapter. Rehabilitation also includes training and treatment, special
schooling (including board and room and travel costs, if necessary),
equipment, and vocational guidance. During rehabilitation, a daily
allowance is paid to provide for the maintenance of the disabled person and his family.
Benefits paid by invalidity insurance are the same as those paid
under old-age and survivors' programs. A full invalidity pension is
paid if the disability amounts to a 66 percent impairment or more,
based on reduction of earning capacity. If the impairment is over 50
percent (or 33 percent in the case of a hardship situation) but less than
66 percent, then a half pension is paid. Also, invalidity pensions are
paid not only in cases of permanent impairment, but also after illnesses
of 360 days followed by at least a 50 percent loss of earning capacity.
The invalidity program is also supplemented by cantonal and
communal assistance programs, as well as by various philanthropic
and private institutions.
The Swiss invalidity program is superior to the U.S. program
in that there are central standards for vocational training and follow-up

that place emphasis on retraining and reemployment, an emphasis
not found in the United States. In the United States vocational rehabilitation is administered by the individual states under federal subvention. There is great variance between the grants, programs, and
degree of emphasis on reemployment. Because there is only a tenuous
tie between the Social Security Administration (federal), which issues
social security disability and supplementary security disabled income
checks, and the state programs of vocational rehabilitation, most social security disability and disablement clients escape the retraining
system unless the client is strongly motivated to want to become employable. Another fault in the system is that many states also have
workers' compensation programs for on-the-job accidents that are
uncoordinated with the federal and vocational rehabilitation programs.
The problem of interpretation of disability in the United States has
caused, first, the disqualification of hundreds of thousands of disabled
clients in mid-1984 who are, in February 1985, again being sought
out by the Social Security Administration for reconsideration of eligibility. The multiplicity of uncoordinated agencies covering the disabled in the United States leaves the field wide open for some clients
to receive regular combined benefits or grants that make retraining
and employment unprofitable, while leaving many disabled without
opportunities to secure information or help in retraining and reemployment. In the United States there are hundreds of high schools for
handicapped children who are allowed to move into permanent disablement "pensions" under the Supplementary Security Income (SSI) provisions of the Social Security Administration without being confronted
with their need to move ahead in schooling and vocational training.
Of all the clients of the U.S. welfare system, it is apparent that the
disabled and handicapped suffer most from the bureaucratization of
federal and state services and the social distances in the mosaic of
separated and disjointed services.

The Swiss invalidity program was included in a study of ten
social security systems conducted by the Union Bank of Switzerland.
The services of each of the systems were categorized according to
two income levels and also categorized according to sickness invalidity versus accident invalidity. At net rates after deduction of taxes,
in income portion received compared with preillness income, these
are as follows (in percent):

Income Category	Sickness Invalidity	Accident Invalidity
Switzerland 1	109.8	109.8
Switzerland 2	116.5	116.5

For category 1, Sweden surpassed Switzerland at 134. 9 percent, Finland at 115. 5 percent, and the United Kingdom at 114. 5 percent under
sickness invalidity. All other countries trailed behind Switzerland at
rates ranging from Canada at 74 percent to West Germany at 97. 3
percent. The United States recorded a 93. 3 percent rate. For category 2 cases, Switzerland's 116. 5 percent was unsurpassed by any of
the other nine countries. The U. S. rate was 76. 7 percent, at the low
end of benefits. For accident invalidity, under category 1, Switzerland surpassed the United States at 93. 3 percent and the Netherlands
at 92. 1 percent. For category 2 cases, Switzerland again surpassed
the United States at 96. 8 percent and the Netherlands at 85. 6 percent.

It is surprising that with such generous invalidity grants in
Switzerland there is not an overuse of this program, as has been observed in Holland. According to Horlick and Skolnik, "An increase
or improvement in invalidity insurance brings an increase in disability
cases" (pp. 113-14), but this has apparently been avoided in Switzerland by careful administration, supervision, and emphasis on retraining, rehabilitation, and reemployment.

THE SUPPLEMENTARY BENEFITS OF THE OLD-
AGE, SURVIVORS', AND INVALIDITY PROGRAMS
(ERGÄNZUNGSLEISTUNGEN PROVISION)

Although the supplementary benefits program represents financial assistance rather than social insurance, it is included in the programs of the social security system. Its location in the social security
system is justified in that it does not serve a population that might be
drawn away from self-sufficiency and employment. The program
came into force on January 1, 1966, with its principal aim to provide
a minimum subsistence level of support for all aged, survivors, or
disabled persons. The supplementary benefits are paid by the cantons,
which in turn receive subsidies in proportion to their financial capacity.
The rates of subsidy range from 10 to 35 percent, depending on cantonal affluence.

As a financial assistance, the supplementary benefits are noncontributory. These benefits nevertheless represent a right guaranteed by law. They are granted on the basis of a means test to all
aged, survivors, or invalids whether Swiss nationals or resident
aliens who have resided in Switzerland for at least 15 years (or 5 years
in the case of refugees). The income levels for eligibility are Fr11,400
for single persons, Fr17,100 for couples, and Fr5,700 for orphans.

In the total income (for supplementary benefits), old-age, survivors', and invalidity pensions are calculated in full. Income from
capital (such as housing rentals from tenants, interest on bonds, and

so forth) is also calculated in full. Of any net capital exceeding
Fr20,000 for single persons, Fr30,000 for couples, and Fr10,000 for
orphans, one-fifteenth of the total is calculated in the threshold total.
Other income, such as pensions for accidents, military pensions,
private pensions, alimony, and family allowances, are only partially
counted. In addition, a number of expenditure items are permitted
to be deducted from the income. These include interest on debts,
liabilities, alimony paid, social-insurance contributions, and life-
insurance premiums. Also, substantial deductions for rent may be
made, up to Fr3,600 for single persons and Fr5,400 for couples. In
addition, in the case of widows and orphans or of invalid fathers and
family members living together or of orphans living together, their
income may be combined, thus making eligibility even more available
(Saxer 1965).

The supplementary benefits program serves to bring aged, survivors', and invalidity benefits up to a basic floor. The payments are
the difference between the adjusted established income level plus the
permitted additional expenses and the actual income received.

A problem in the supplementary benefits program is reported
by Janssen and Müller (1982). It appears that many people in rural
areas who are entitled to these benefits do not claim them, perhaps
because of embarrassment or stigma. Janssen and Müller recommended that ways should be sought to make these benefits more accessible. It should be noted that most community tax offices in Switzerland check the income tax reports required of all residents. When
income is below the Fr12,000 mark, and when the person is aged or
widowed or invalid, the local clerks seek to encourage the tax filer
to apply for supplementary benefits. The problem is that many persons who are eligible for supplementary benefits are reluctant to accept an unearned income because acceptance of it might represent a
"loss of face." This income-stratification problem will be discussed
later in this chapter.

SICKNESS INSURANCE AND ACCIDENT INSURANCE

Along with military-injury insurance, sickness insurance is the
oldest social insurance in Switzerland. The sickness-insurance funds
originally were a part of the services provided by the old fraternities
or guilds that were mentioned previously as a factor in Swiss cohesion
and consensus. Some of these fraternities were established in the
Middle Ages, when merchants and businessmen sought mutual protection of their interests and organizational effectiveness. The economic
costs of sickness became one of the concerns of the groups that sought
to utilize mutual-benefit approaches in the resolution of their problems.

Many of the businesses were designated corporations in the fraternities, and employees were designated companions. Soon the companions began to form their own mutual-benefit associations, including sickness-benefit associations.

The Federal Act of 1911 established general regulations and standards for sickness and accident insurance associations (Charles 1983). All such associations that met the federal standards were made eligible for federal subsidies.

The sickness funds are administered on a regional basis, with 600 approved funds. These funds are required to operate on the mutual-benefit principle (nonprofit), to provide equal treatment to all insured persons, to provide at least the minimum standard benefits for their members, to provide adequate financial security necessary to meet their commitments, and to accept periodic audits and control by the Federal Social Insurance Bureau.

The sickness-insurance coverage is individual. The insurance is voluntary, although cantons are empowered to declare a particular insurance compulsory for any or all population categories in the canton, and some have done so. Every person who is a resident or gainfully employed may become a member of the sickness-insurance fund of his or her choice. Admission may not be refused on grounds of health or pregnancy, but specific existing medical conditions may be excluded. Funds may set a maximum age limit for admission (for the purpose of encouraging early and lifelong enrollment).

The law sets only minimum benefits, although many funds provide higher benefits based on higher corresponding contributions. Benefits include in-kind services as well as cash payments. The former include doctors' fees, recognized treatments, medicines, drugs, tests, analyses, and ancillary services (such as chiropractic treatment). A copayment of 10 percent for treatments is required of all insured, but there is no copayment required for hospitalization costs. There is also a fixed franchise fee equivalent to a deductible each year.

The minimum sickness allowance is still set at Fr2 per day, as in 1911, but most sickness funds offer packages of services of Fr50 to Fr180 per day, requiring corresponding prepaid contributions.

Schweizer (1984) lists the sickness-insurance provisions for a family with three children and an income of Fr24,000 per annum (US$10,572). In such a family, the father's illness or disability would be covered by insurance income amounting to 95 percent of his regular income (Fr23,000, or US$10,132).

The family's contribution to sickness insurance depends on the insurance coverage purchased, the level of benefits, geographical factors, and the age of the insured. Women, whose total sickness expenditures have been found to exceed men's by 50 to 60 percent in Switzerland, pay sickness contributions of no more than 10 percent above those of men.

The purpose of the copayment and the deductible is to discourage overuse of facilities, as well as to reduce expenses of the sickness funds.

A partial revision of the sickness-fund system is presently under way. The need for better coverage for various categories of population (especially pregnant women) and control of the continued rise of insurance costs have occasioned discussion of proposed changes in the law. These include coverage of hospital services beyond the current limit of 720 days within each 900-day period, extension of benefits to cover home treatment and preventative medical services (including annual physical examinations), extension of services to cover general dentistry, extension of daily allowances for pregnant women from 10 to 16 weeks, contributions to help cover home expenses for new mothers and babies, and the introduction of compulsory insurance for loss of earnings. Also to be considered are allowances for total incapacity up to 540 days (with 80 percent of previous income). A number of health-cost controls are also being considered, including improved control of agreements between the sickness funds and medical vendors and an increase of deductibles to 20 percent instead of 10 percent.

An interesting Swiss pattern used by some of the sickness funds is to make payments only to the insured, who in turn pays the physician or other medical vendor. This method of payment is useful in bringing the actual expenditures to the attention of the insured, who then can be motivated to check the items for accuracy or necessity. The patient's motivation derives from the copayment, as well as from the Swiss propensity to examine all expenses carefully. This pattern may in part explain the much lower increase in health-care costs experienced in Switzerland when compared with other Western nations.*

Estimates of the extent of coverage have been listed as over 90 percent in 1979 (Büchi et al.). Estimates of over 94 percent were reported during interviews by the author with experts in 1983, and it became clear that all uninsured persons are provided for by local welfare offices.† Federal contributions to the funds amount to about 18 percent of total expenses.

Persons who are not covered by the sickness-insurance program and who do not have adequate resources have their medical and hospi-

---

*Another factor used by Swiss commercial health-insurance companies to control health-care costs is their administrative limits on client-induced risks. An example of this is the lowered health and accident benefits granted in automobile accidents where clients have been found not to have used safety belts required in the policy.

†Except, of course, the uninsured affluents, who prefer to provide their own private coverage.

tal costs paid by the community welfare offices. Where persons are covered but cannot pay the copyaments, these are provided on an as-needed basis by the welfare offices. Sickness-insurance premiums carried by persons on the supplementary benefits program are permitted to be discounted from their income listing (making payment possible from their resources without losing any of their supplementary benefits).

Accident insurance is regulated by the Federal Act of 1911 and is compulsory only for employed persons in industry, handicrafts, and agriculture. It is administered by the Swiss National Accident Insurance Institute (SUVA) on a mutual-benefit basis under the supervision of the Federal Social Insurance Bureau. Under the Federal Act of March 1981, after January 1, 1984, the program was expanded to provide compulsory coverage for all employed persons. Self-employed are now included on a voluntary basis at the same rates, either through SUVA or through private insurance companies or recognized sickness funds.

Under the accident-insurance program, a number of risks are covered. These include medical care; daily allowances amounting to 80 percent of insured salary, starting on the third day after the accident; 80 percent of the insured salary for total invalidity; and survivors' pensions of 35 percent of salary for widows and 15 percent for each child still in educational or occupational training, up to a total of 70 percent of the prior salary of the insured. Where the beneficiaries are also eligible for IV, then the total benefits may not exceed 90 percent of the insured income.

An important aspect of the accident-insurance program is the program of accident prevention. Employers are required to undertake a variety of preventative measures to eliminate accident or disease factors in their plants and seek the safest conditions possible. In this, SUVA and accident-insurance companies are active in working with employers, much more so than is reported in other Western nations.

UNEMPLOYMENT COMPENSATION

Tschudi (1978) reports on unemployment-insurance programs prior to 1977, which proved unsatisfactory in the period during the mid-1970 depression when only a third of the active population was covered. Beginning April 1977, all employees are covered by compulsory unemployment compensation. Premiums are collected by the AHV administration. The contributions, totaling 0.3 percent of wages, are paid by employer and employee in equal amounts. The funds collected by the AHV are paid out to the recognized private or cantonal unemployment-insurance funds as required. Daily unemployment payments amount to 70 percent of the insured income, up to a maximum of Fr78,000 per annum (Fr6,500 per month), with a supplement for each dependent of Fr190 per day. The number of insured days is limited to 155, with an extension available for older employees or in regions affected by a recession.

The purpose of the program is to protect involuntarily unemployed persons against a major loss of earnings. At the same time, the program seeks to prevent workers from "getting fixed in unemployment" (Charles 1983, p. 30). This is accomplished by rules and measures to facilitate and encourage the resumption of work. Arrangements and resources are available for special vocational training and retraining, payment of subsidies to encourage geographical mobility, and professional and intensive assistance in reemployment.

Schweizer (1984) cites unemployment payments of Fr1,650 (US$726) per month for an employee whose regular income was Fr2,000 (US$881). This is an 83 percent rate. For employees earning Fr5,000 (US$2,202) per month, the monthly benefit is Fr2,550 (US$1,123), approximately 50 percent. For persons unemployed longer than 150 days, only community public welfare is available on the basis of a means test and at a lower rate than unemployment compensation.

FAMILY ALLOWANCES

Tschudi (1978) discusses the difficulty in arriving at a consensus on what should be the scope and extent of family allowances. To what extent should employees be paid differentially because they have more children (rather than because they are more competent or productive)? Because of this lack of consensus on the subject, it has been impossible to arrive at a national set of standards for children's allowances.

All employed persons in each canton are required to belong to a family-allowance equalization fund, to which contributions of 1.4 percent to 3.5 percent are paid. Cantons pay from Fr70 to Fr150 monthly per child. There are also allowances for vocational training, ranging from Fr100 to Fr180 monthly per child. The birth allowances range from Fr300 to Fr500. Payments are made for all legitimate, illegitimate, step, and foster children up to the age of 18 years (25 years in instances where further study or training is undertaken). Because the programs are operated on a local level and because the allowances are limited to amounts not approaching take-home pay, the program has none of the disincentives to employment found in the U.S. Aid to Families with Dependent Children (AFDC) programs. (A possible exception to the nondisincentive effect of Swiss family allowances occurs in instances where six or more dependent children are in a family. In such cases, the family allowance income does permit employment avoidance.)

It should also be noted that deductions on income taxes also serve to provide economic advantages to families with children. To examine the impact of the family allowances and tax deductions, Schweizer (1984, pp. 11-12) presents a comparison of the economic situation between a childless couple and a family with one, two, and more children. If both couples have a base monthly income level of Fr2,000 (US$881), the net income, aside from family allowances and after deductions for taxes and social security contributions, rises with one child to 103.5 percent, with two children to 108.2 percent, and with six children to 122.6 percent.

It should be noted that the social security system and other social insurances are not viewed as final and constant. The programs are constantly studied at all levels of government and revised as deemed necessary.

The financial experience of Swiss social insurance in 1980 is listed in Tables 3.3-3.6, which follow.

Parallel to an examination of the social security system is the necessity to study the distribution of income in Switzerland.

There has been considerable discussion of the distribution of income in Switzerland. Schweizer (1980) examined the economic condition of pensioners in Switzerland and found that most pensioners, even though they may view themselves as of modest means, really have sizable assets. Many of these aged and handicapped have not taken into consideration assets available to them over and beyond their income. There are, however, some one hundred thousand to two hundred thousand people in the AHV or IV pension group or pension-eligible group who have a level of income and assets that place them as potentially at risk of poverty. They are mostly women of low educational achievement, with little or no career base. Most of these women are not married, having either never married or been early divorced or widowed. These women were brought up in families that did not expect them to work, and many were raised in protected families only to be stranded in later life after the families were no longer around. Many of them might be out of poverty if they were informed of the supplementary benefits and were willing to apply for them. Thus, it seems that the problem of this population is primarily one of attitude rather than of problems within the social security system. In the long run, however, this may prove to be a generational problem based on past patterns of childrearing no longer in existence in Switzerland. With the current Swiss program emphasis on educational and vocational training for females as well as for males, it seems apparent that this problem will decline.

TABLE 3.3

The Swiss Social-Insurance System: Income, 1980
(in millions of Swiss francs)

Income	From Insured and Employees	From Employers	Confederation Participation	Income from Capital and Other	Communal Contributions	Total Receipts
Sickness insurance	3,877.9	57.6	912.8	184.1	304.8	5,337.2
Unemployment insurance	214.3	214.3	—	45.2	—	473.8
Employment injury insurance	576.6	577.7	—	424.8	—	1,579.1
Old-age and survivors' insurance	4,752.8	3,873.0	1,394.3	339.1	536.3	10,895.5
Invalidity insurance	570.3	464.7	806.9	—	269.0	2,111.3
Complimentary benefits*	—	—	215.1	6.7	195.5	421.3
Cantonal old-age, invalidity, and survivors' insurance	1.4	—	—	7.1	0.1	8.6
Total	9,993.3	5,187.3	3,329.1	1,007.4	1,309.7	20,826.8

*Represents supplementary (noncontributory) program of old-age, survivors' benefits, and invalidity insurance for aged, infirm, or survivors with per capita income of less than Fr12,000 (after various deductions).

Source: International Labour Office, The Cost of Social Security: Basic Tables (Geneva: International Labour Office, 1985).

TABLE 3.4

The Swiss Social Insurance System: Expenses, 1980
(in millions of Swiss francs)

Expenses	Medical Care and Other Benefits in Kind	Cash Benefits	Total Benefits Distributed	Administrative	Transfers to Other Schemes and Other Expenditures	Total Expenditures
Sickness insurance	3,951.5	529.6	4,481.5	430.9	352.1	5,264.5
Unemployment insurance	—	103.9	103.9	19.8	29.6	153.3
Employment-injury insurance	248.5	871.5	1,120.0	133.7	1.9	1,255.6
Old-age and survivors' insurance	7.5[a]	10,578.4	10,585.9	48.4	91.3	10,725.6
Invalidity insurance	346.9[b]	1,440.3	1,787.2	13.5	351.0	2,151.7
Complimentary benefits[c]	—	421.3	421.3	—	—	421.3
Cantonal old-age, invalidity, and survivors' insurance	—	9.8	9.8	0.8	—	10.6
Total	4,554.8[d]	13,954.8	18,509.6	647.1	825.9	19,982.6

[a] Benefit in kind other than medical.
[b] Medical benefits, 315.5; other benefits, 31.5.
[c] Represents supplementary (noncontributory) program of old-age, survivors' benefits, and invalidity insurance for aged, infirm, or survivors with per capita income of less than Fr12,000 (after various deductions).
[d] Medical benefits, 4,515.8; other benefits, 39.0.

Source: International Labour Office, The Cost of Social Security: Basic Tables (Geneva: International Labour Office, 1985).

TABLE 3.5

Swiss Special Insurance Systems and Allowances Income: Income, 1980
(in millions of Swiss francs)

Income	Contributions from Insured Persons and Employees	Contributions from Employers	Confederation Participation	Income from Capital and Other Receipts	Communal Contributions	Total Receipts
Family allowances, agricultural workers	—	6.4	41.5	—	20.8	68.7
Military insurance	—	—	173.5	—	—	173.5
Federal insurance fund	179.4	473.0	—	256.0	—	908.4
Federal railway pension fund	58.7	189.6	—	97.1	—	345.4
Cantonal staff pension funds	297.6	672.8	—	463.4	—	1,433.8
Public health services	—	—	8.8	—	2,997.7	3,459.8
Total	535.7	1,341.8	223.8	816.5	3,018.5	6,389.6

Source: International Labour Office, The Cost of Social Security: Basic Tables (Geneva: International Labour Office, 1985).

TABLE 3.6

Swiss Special Insurance Systems and Allowances Income: Expenses, 1980
(in millions of Swiss francs)

Expenses	Medical Care	Cash Benefits	Total Benefits	Adminis- trative	Other Expenditures	Total
Family allowances, agricultural workers	—	66.9	66.9	1.8	—	68.7
Military insurance	55.5	108.2	163.7	9.8	—	173.5
Federal insurance fund	—	443.4	443.4	—	25.6	469.0
Federal railway pension fund	—	215.1	215.1	—	8.3	223.4
Cantonal staff pension funds	—	607.1	607.1	—	91.9	699.0
Public health services	3,459.8	—	3,459.8	—	—	3,459.8
Total	3,515.3	1,440.7	3,526.7	11.6	125.8	5,093.4

Source: International Labour Office, The Cost of Social Security: Basic Tables (Geneva: International Labour Office, 1985).

88

A discussion of Schweizer's findings was reported in the Neue Zürcher Zeitung of September 18, 1980, and was criticized there also by Urs Ernst in May 1983. Ernst used the studies of Gilliand (University of Lausanne) and Lüthi (University of Freiburg) to counter Schweizer's findings. Unlike Schweizer's finding of an average annual income of Fr29,400 in 1976, these scholars, utilizing a "corrected" base, found an average annual income of Fr26,700 in that year. The difference of Fr2,700 became the basis for disagreement as to the level of income defining poverty. Ernst believed that this difference led to an underestimation of the income differences between pensioners and the working population. He indicated, however, that the critics overemphasized the error in the Schweizer study, and he believed that differences in income had been overdrawn. Ernst concluded that in 1978, some 20 percent of all Swiss required to pay taxes and those on AHV pensions had a net income, before tax deductions, of Fr12,000 or less. This low-pension group had an average annual income of Fr10,300. The median net income in 1978 in Switzerland was Fr22,900. The top 10 percent of all pensioners draw an average income, before taxes, of Fr95,900. The average income for all pensioners that year was listed at Fr46,100.

Compared with other Western nations, the gap between rich and poor pensioners in Switzerland is not great. Whether one calculates these income levels by the Schweizer or Gilliand and Lüthi methods, one needs to realize that the pensioners at the lower level of the income scale have available not only the supplementary benefits, but also the aid of the local welfare offices and a variety of voluntary services. From interviews with the Zurich, Bern, Basel, and Geneva welfare offices, it was learned that sizable welfare grants are provided to the aged and ill, particularly in the form of support for specific needs, as reported by them. It should also be remembered that many of the poorer pensioners are less dependent upon government aid because they are still in touch with their children and other relatives (unlike the situation in the United States) and are often provided with regular and substantial aid by relatives. It should also be noted that the Gemeinschaft pattern of community life provides considerable neighborly aid in Switzerland, unlike the situation in other lands.

Another view of the income distribution patterns of Switzerland is provided by Hoffmann-Novotny et al. (1978), who present income distribution figures for 1969/70 in Switzerland in data and graph form. They report that half of the married net-income earners received between Fr15,000 and Fr25,000; one-fourth earned between Fr25,000 and Fr40,000 net. Of all income, 70 percent goes to the 90 percent of the population with annual net incomes under Fr40,000. The remaining 30 percent is distributed among those with net incomes higher than Fr40,000. At the lowest end of the scale are the 13 percent of income receivers with less than Fr9,000 net annually; at the other end of the scale, there are 0.02 percent (1,185 people) with annual net incomes of more than one million francs.

It should be noted that these figures include beginning workers, apprentices, part-time workers, women, part-time students, and invalids at home who receive some income. It is obvious that many of the people at the lower end of the income scale are resident in homes where others (primarily husbands or fathers) bring in the bulk of the family income. The fact that 90 percent of the population receive 70 percent of the income makes it clear that the Swiss are primarily a middle-class country. To us, at least, the income distribution pattern in Switzerland shows a restrained set of income differences. Such income differentials as do exist are to be expected in the light of an economy where most people are motivated and productive.

Obviously, the social-insurance systems and the income distribution patterns of Switzerland serve to keep the population productive, employed, and basically self-sufficient. This cannot be claimed for many of the other Western nations.

REFERENCES

Büchi, O., A. Inglin, A. W. Stahel, and Peter Tschümperlin. 1979. "Die Organisation des Sozialwesens in der Schweiz" [The organization of social policy in Switzerland]. Swiss section of working papers prepared for the three-country student assembly (West Germany, Austria, and Switzerland) held in Frankfurt, November 19-23, 1979, under the auspices of the Max Planck Institute for Foreign and International Justice at Munich, pp. 1-14.

Charles, Jean-François. 1984. Social Security in Switzerland: Main Features of the Schemes and Current Problems. Report on Swiss social insurance made to the Twenty-first General Assembly of the International Social Security Association, Geneva, October 3-13, 1983, p. 181. Bern: Federal Office of Social Security.

_____. 1983. Social Security in Switzerland: Main Features of the Schemes and Current Problems. Bern: Federal Office of Social Security.

"Die Wirtschaftliche Lage der Rentner." Neue Zürcher Zeitung, September 17, 1980, p. 33.

Ernst, Urs. 1983. "Zur Einkommenslage der Schweizer Rentner." Neue Zürcher Zeitung, May 26, 1983, p. 33.

Gruner, Erich. 1982. Personal correspondence with the author, March 26.

Hauser, Jürg A., Franz Bluntsch, Rudolf Hohn, Elizabeth Monig, and Lorenz Wolfensberger. 1980. Empirische Aspekts der Fürsorge bedurftigkeit: Am Beispiel des Kantons Zürich [Empirical aspects of public welfare neediness: an example of the canton of Zurich]. Bern: Paul Haupt.

Hoffmann-Novotny, Hans-Joachim, Robert Blancpain, François Hopflinger, Martin Killias, Matthias Peters, and Peter Zeugin. 1978. Almanach der Schweiz. Bern: Peter Lang.

Horlick, Max, and Alfred M. Skolnik. 1978. Mandating Private Pensions: A Study of European Experience. Washington, D. C.: U. S. Department of Health, Education and Welfare, Office of Research Statistics research report no. 51.

International Labour Office. 1985. The Cost of Social Security: Basic Tables. Geneva: International Labour Office.

Janssen, Martin C., and Heinz H. Müller. 1982. "Switzerland." In The World Crisis in Social Security, edited by Jean-Jacques Rosa, Richard Hemming, John A. Kay, Oronato Castellino, Noriyuki Takayama, Ingemar Stahl, Martin C. Jannsen, Heinz H. Müller, Sherwin Rosen, Karl Heinz Jüttemeier, and Hans-Georg Petersen, pp. 121-47. San Francisco: Institute for Contemporary Studies.

_____. 1981. Social Security in Switzerland: Provision for Old Age and Survivors. Institute for Empirical Research in Economics, report no 8102. Zurich: University of Zurich.

Lüscher, Kurt K., Vernea Ritter, and Peter Gross. 1983. Early Child Care in Switzerland. London: Gordon & Breach.

Saxer, Arnold. 1965. Social Security in Switzerland. Bern: Paul Haupt.

Schulte, B. 1980. "Reforms of Social Security in Europe in the Periods 1965-1975 and 1975-1980. Colloquy in Perugia. Munich: Max Planck Institute for Foreign and International Justice, pp. 6, 8, 19-20, 31, 32.

Schweizer, Willy. 1984. "The Social Security System of Switzerland." Paper presented as class lecture at various universities in Taiwan, Hong Kong, and Singapore.

_____. 1980. Die Wirtschaftliche Lage der Rentner in der Schweiz. Bern: Paul Haupt.

Social Services of the City of Bern. 1981. Client Groups and Situational Problems, Comments, and Statistics for 1980. Addendum to Administrative Report. Bern: Social Services of the City of Bern.

Swiss Conference for Public Welfare. 1980. "Standards for the Apportionment of Material Assistance." Paper presented at the Swiss Conference for Public Welfare, Bern.

Switzerland, Federal Office of Social Security. 1982. Social Security in Switzerland: Summary of the Main Benefits. Bern: Federal Office of Social Security.

Tschudi, Hans Peter. 1978. "Social Security." In Modern Switzerland, edited by J. Murray Luck, Hugo Aebi, Joseph von Ah, Lukas Burckhardt, Erich Gruner, and Hans Haug, pp. 199-212. Palo Alto, Calif.: Society for the Promotion of Science and Scholarship.

Tuggener, H. 1981. Personal correspondence with the author, December 3.

Union Bank of Switzerland. 1977. Social Security in Ten Industrial Nations. Zurich: Union Bank of Switzerland, Economic Research Department.

Urner, P. 1980. Principles of Public Welfare Law in Zurich. Zurich: Agency of the City of Zurich.

U.S., Department of Health, Education and Welfare. 1976. Social Security Programs throughout the World, 1975. Washington, D.C.: Government Printing Office. (Updated to 1981 by Jean Charles, Office of Federal Social Security, Bern.)

Welfare Agency of the City of Bern. 1981. Verwaltungsbericht, 1980. Bern: Social Services of the City of Bern.

Welfare Agency of the City of Zurich. 1979. Annual Report. Zurich: Welfare Agency of the City of Zurich.

_____. 1978. Annual Report. Zurich: Welfare Agency of the City of Zurich.

4

SWISS IMMIGRATION POLICY AS A CONSTRAINT ON WELFARE DEPENDENCY

To most analysts, the presence or availability of unlimited labor resources in a finite employment market can be expected to drive wages down, to put pressure on public social services, and to lead to high rates of unemployment. If society has too many people to support and only finite resources, problems of poverty can be expected. It was for this reason that, for many decades, second and third sons in Swiss families had to leave home and "seek their fortune" in other lands as settlers or mercenaries. It was for this reason that those second or third sons who married and stayed on were shunned and isolated in the Swiss society. And it was for this reason that immigration was carefully limited. A society that had experienced hunger in the past and was surrounded by nations full of other unemployed and hungry people could be expected to focus on solving its own economic problems. A small society seeking to provide employment for its own people can hardly leave its borders open to all, no more than a housekeeper in winter can warm the house while leaving all the windows and doors open. Many small nations have seen the relationship between limited resources and open doors. Norway, Denmark, Sweden, and other nations, in their hunger to improve the quality of life for their populations, have maintained rigid controls for decades.

The Swiss method of immigration control is somewhat different and perhaps somewhat more rational than the immigration processes of other countries. In other lands, people who wish to be admitted make application, and only certain categories, nationalities, and occupations are considered. Some are accepted and many others wait years and may not be accepted. Those who are admitted are permitted to stay, even if they fail to make a satisfactory adjustment. Those who fail to find jobs live out their lives without employment or in some other social difficulty. Having cut their ties to their home country, they usually cannot return.

The Swiss method is different. Persons wanting to get a job in Switzerland in order to try out life and employment in that land may come in as seasonal workers or on a one-year contract. If they do well on the job or find a more attractive job, it is possible to remain on a year-by-year basis. During this period, if both workers and families are law-abiding, and if their work skills are improved or even merely maintained, they are permitted to stay on. After five years on annual contracts, permits are granted on a two-year basis. After ten years, the workers apply for resident status, under which they are no longer required to secure a contract, and may remain indefinitely. The foreign workers must serve the first year without their families. During the first ten years, they must have permission to move to another canton or to change jobs. The reason for this is that foreign workers are brought into Switzerland to fill vacant employment positions, just as Swiss workers are required to secure work permits to fill jobs in France, Italy, or other countries. Aside from these restrictions and up to the time of securing residency, the foreign workers and their families have all the rights that others have. They can join trade unions and political parties and participate in the workers' committees in industrial plants (Kümmerly and Frey 1984, p. 25); they are entitled to all of the social-insurance benefits earned; they have all the rights of other residents; and when their children reach the age of 18, they may also apply for citizenship. Foreign workers can return to their home countries during vacations. On federal issues particularly, they have the right to speak out without hindrance; and if they are in their home country or elsewhere, they can receive social-insurance benefits by mail.

After earning the right to stay on as permanent residents, foreign workers may choose a community where they can work toward citizenship. Each year many thousands are accepted.

This system of retaining the immigrants who are competent on the job, law-abiding, and satisfactorily settled and of setting a probation period for all immigrants is probably more humane in the long run for both immigrants and society.

In recent years, some critics of Swiss immigration policy, from within and without Switzerland (including some who live in lands with even more restrictive immigration policies), have claimed that Switzerland "exports" its unemployment. This criticism originally was voiced during the post-oil-crisis recession, when some 200,000 annual contract workers were refused renewals because of a lack of available employment. This number is relatively small when one considers that Switzerland has the highest proportion of foreign workers in the Western world (14.4 percent). Thus, every seventh person in Switzerland (in addition to tourists and visitors) is a foreigner.

As of August 31, 1984, foreign population in Switzerland is as follows:

	Total Number	Number Gainfully Employed
On seasonal contracts	100,753	100,753
On annual contracts	197,194	117,263
Resident aliens	727,473	414,853
Total	1,025,420	632,869

The 392,551 difference between total number of such foreigners and the number of gainfully employed foreigners represents persons between jobs or family members. It should also be noted that a sizable number of the annual contract workers are employees of international organizations in Geneva and are listed among the contract workers.

In addition to the above foreign population, there are also about 105,000 foreigners who live in West Germany, Italy, Austria, and France who daily cross the borders to work on regular jobs in Switzerland. These are the so-called Grenzegängers, many of whom occupy responsible positions in Swiss businesses and institutions and who are granted continuing border-crossing and work permits in Switzerland because their services are essential in the Swiss institutions where they are employed. Ricq (1982a) suggests that these "border crossers" into Switzerland are offset, at least in part, by others who cross daily from Switzerland into other neighboring lands.

In addition to the seasonal workers, the contract workers, the resident foreign workers, and the border-crossers, Switzerland has recently been experiencing an increase in requests for asylum. Until the end of the 1970s, there were about 1,000 such requests per year. By 1978 the annual number rose to 1,389, in 1979 to 1,882, in 1980 to 3,020, and in 1981 to 4,226. For 1982 the estimate approached 1,982 (the latest available figure, see Hadorn 1983). Hadorn, of the Federal Bureau of Immigration Police, expects the problem to increase with the coming years. These applications are made to the cantonal authorities, which are required to accept them conditionally regardless of the degree of validity of the request. During the conditional period, the federal agency reimburses the cantons for the welfare maintenance costs of refugees. The applications are then processed and decided upon by the federal office, and meanwhile, the refugees and their families are permitted to accept employment. If the applications are refused, then the applicants can be deported (only in extreme cases) or, most frequently, they are restricted to a stay of two to three years. Many of these applications drag on for years, by which time their passports may have expired and the refugees, having no travel documents, are kept on in Switzerland at public expense or permitted to accept employment and remain indefinitely.

Because many of the refugees come from countries with cultures vastly different from the Swiss culture, the process of determining the validity of the applications and the process of aiding the refugees to settle in the communities has proved quite difficult.

Any discussion of foreign workers in Switzerland requires examination of the proportion of native to foreign workers. Considering that there are 3,012,700 gainfully employed people in Switzerland, 737,869 of which are employed foreigners, that means that about one out of every four workers is a foreigner. This is a condition that few other nations would find acceptable. Imagine, for example, that every fourth worker in the United States were a foreigner and that, in addition, half of these foreigners had their families with them. This would mean a foreign labor force of about 26 million workers plus about 13 million foreign family members—a total of about 40 million people. How easily would Americans adjust to such an atmosphere? And yet, the Swiss calmly go about their affairs, operate their businesses and organizations, and maintain a functioning society with a minimum of fuss or disturbance.

To put the situation in another format, imagine a group of families comprising 65 people who live on a small island in the middle of a river. The families decide that they need to build some houses, but they do not have enough workers. So they invite workers to come to the island to help in the building. They make a contract with these hired workers to provide them with housing, food, and protection against injury, as well as payment. When the job is finished, most of the workers go home. A few find other jobs on the island and make new contracts for the new jobs. If some of the workers wanted to stay but no one would hire them because their work was slow or for any other reason, and if the island were crowded, would these families be criticized for asking these unneeded workers to leave? One can no more expect Switzerland to become a haven for nonworkers than any one of us would expect the "man who came to dinner" in our house to stay on forever. In the final analysis, a work contract is not an immigration contract unless that has been specified in the beginning.

The conditions of the employment recession of 1974-75 in Switzerland make the problem quite clear. In that period, out of a total work force of about 3,203,000, a total of 340,000 jobs were lost. In the necessary cutbacks, some 230,000 foreigners did not have their work contracts renewed. At the same time, about 110,000 other workers lost their jobs. The job losses were rather equally spread: 19 percent in construction, 13 percent in textiles, 17.4 percent in the watchmaking industry, 12.7 percent in clothing, and 12.5 percent in wood and furniture industries. In the case of the seasonal and contract workers, it is in error to describe such workers as dismissed. Rather, it is appropriate to describe them as workers for whom there is no longer a position and who were not rehired. From the point of view of compassion, it would be cruel to promise workers from another land a job indefinitely and then to send them away when there is no work. To keep someone around waiting for work is also hardly humane. A factory that keeps people on when there is no work for them is destined to go bankrupt. The same applies to a nation that keeps visitors on who have no jobs.

No nation can become the employer of last resort for all of its neighbors, and to speak of "exported unemployment" is equivalent to requiring of Switzerland a policy that no other nation would be ready to undertake.

It is worthwhile to examine the immigration patterns in other European countries. According to Schmid (1980), there were about eight million foreign workers in Western Europe and the United Kingdom in 1980, accompanied by four million to five million dependents. More than 80 percent of these migrants were resident in West Germany, the United Kingdom, France, and Switzerland. The remaining 20 percent were in the Netherlands, Belgium, Sweden, and Austria. All of the receiving nations had experienced rapid economic growth since World War II. Other peripheral countries (Greece, Italy, Portugal, Spain, Turkey, Yugoslavia, Finland, and Ireland) had experienced serious unemployment and underemployment deriving from stagnant economies, improved health conditions, and extended life expectancies, which were not offset by birthrate controls. These eight countries provided 75 percent of the foreign labor employed in Western Europe, the balance coming from France and England. Switzerland, as far back as 1914, depended upon foreigners, excluding border crossers, for 15.4 percent of its labor force. This can be compared, in 1975, with foreign labor components of 11 percent in France, 10 percent in Germany, and 8 percent in the United Kingdom.

Until recent years, the foreign-labor force in Switzerland came from neighboring countries—West Germany, Italy, Austria, and France. In 1910 these countries accounted for over 95 percent of the Swiss foreign-labor component. By 1970 Switzerland experienced a decrease in the proportion of French and West German foreign workers and an increase in Italians. By 1977 the Italians made up about 50 percent of the Swiss foreign-labor force. There was also another gradual change in the makeup of the foreign labor force: more workers were manual wage earners (from 58 percent in 1941 to 89 percent in 1970). There were over 200,000 foreign seasonal workers in the Swiss labor force by 1970. Foreign workers became highly concentrated in certain industries on jobs that were usually rejected by Swiss citizens. Foreign men, able and willing to move around, were most likely to go into construction work, and foreign women were most likely to work in the textile industry. This does not mean that foreign workers stayed indefinitely in these industries. Many used these industries as entry-level jobs, which they then left when they found more remunerative or attractive employment. Yet another way in which the immigrant population moved upward in the society was through the advancement of some of their children. In 1983, for example, over 400,000 immigrants were on residence permits (almost half of the foreign population). These children in foreign families benefited from the educational and vocational training provided in Switzerland to all residents. As long as the parents were hardworking and aspiring for upward mobility, many of the children moved ahead in terms of education, training, and employment advancement. Thus, the process of screening out the serious workers for continuance as guest workers means that the foreign children who remain have the kind of parents who are apt to reinforce academic and vocational advancement for the children.

Of course, many of the foreign workers take jobs that some would describe as "dirty work." This is acceptable in a democratic society as long as there are permeability-mechanisms in the society by which people can move upward. In many other countries with high levels of immigration, the foreign workers may be provided with more readily obtained residency status but may, in turn, find that their equality is primarily for a share in unemployment compensation and the dole. Of course, in many of the nations with a sizable proportion of foreign workers, the labor unions take an ambiguous position in regard to them. This position is usually officially in support of the foreign workers, but unofficially, there is resistance to foreign workers who may work harder than the natives or who may be at the roots of native political fears of being displaced by foreigners. Unlike other countries where there are constant confrontations between labor and capital, in Switzerland there is a consensus of labor-capital cooperation, which is probably sufficient to prevent such displacement fears of native workers. There has been a level of unwillingness of foreign workers to join Swiss unions, but with the continued retention of many foreign workers as permanent residents, this trend is on the wane. There has been a continued effort by most labor unions in Switzerland to integrate foreign workers into the labor force and the trade union movement, along with constant pressure for foreign workers to be given equal pay and working conditions. Kümmerly and Frey (1984) report that in April 1981, a "togetherness" proposition was placed before the Swiss electorate "by thousands of xenophiles" (p. 25). This proposition sought equal status (economic, social, and legal) for foreigners at the moment of their arrival and abolition of the seasonal-worker status. This proposition was rejected by the electorate and the cantons. From interviews with Swiss scholars as well as some of the political leaders, it has been learned that the primary objections to the proposition derived from the issues of eliminating the seasonal-worker status and the requirements of residency permits. The fear of being flooded by thousands of instant residents was a realistic one, and it was the belief of many of those interviewed that the plebescite might have passed if it was limited to the granting of equal rights to foreigners with resident status. There has been action since that time in the Federal Council to prepare a new law proposing a progressive consolidation of foreigners' legal status.

Neither Switzerland nor the other labor-importing countries of Western Europe view themselves as a nation of immigrants in the same way that Americans have viewed their country as a permanent landing for immigrants.* Most of the Western European countries view their foreign workers as <u>Gastarbeiters</u> ("guest workers") rather than immigrants. Many of these countries have recently sought to steer these workers back to their home countries by various means. West Germany provides a sizable bonus payment to guest workers who agree to leave. Only Switzerland is in the process of getting down to the business of absorbing its resident foreign workers, at the same time preventing others from entering unless they clearly understand that their entry is on a contractual time-limited basis.†

---

*In the years 1941-84, the United States admitted 12,026,000 immigrants. This amounts to about 5 percent of the U.S. population. During those same years, Switzerland has admitted (and retained) about 14 percent of its population.

†It should also be noted that Swiss immigration and guest-worker policies are strongly influenced by cantonal concerns, particularly in those cantons that border on foreign states. Unlike the situation in many other countries where immigration and guest-worker policies are operated as a part of national foreign policy, in Switzerland the border cantons participate actively in discussions and negotiations with bordering countries (Ricq 1982b). On the matter of immigrant absorption, there is a need for more active involvement of the Swiss communities in providing mechanisms by which new immigrants can be helped to learn the necessary "mainstream languages" (either German, French, or Italian, as required for their commune of residence). Similar mechanisms for learning the appropriate social patterns are necessary. There is also a scarcity of "catch-up" and "second-chance" schooling opportunities in the communities, which could be utilized for integration of adult immigrants as well as for

Many of the seasonal and guest workers in Switzerland have no
initial intention to settle in Switzerland. The author interviewed one
young guest worker who had just graduated from a hotel school in
Wales. He had come to Switzerland on a one-year contract with the
Swiss National Railways, which operate restaurants in many of the
terminals. His assignment was to work half days in the kitchen,
learning how to cook, making up menus, and so forth, and the other
half days working with the purchasing agent and management of the
restaurants. He works a 44-hour week and receives full board and
room in an employees' hotel, plus a train pass and some Fr350 per
week. After the completion of the year, he will be granted a certifi-
cate of completion for the apprenticeship and then can apply for employment to other hotels or restaurants in Switzerland, England, or
elsewhere. The young man viewed this contract as an opportunity
and saw it as "a much better deal" than to stay home in Wales and
depend on unemployment compensation and the dole. I asked him why
more Swiss youth did not apply for these apprenticeships with the
railway, and he answered that many preferred the apprenticeships
offered by four-star hotels (something that was not open to him in
Wales).

Schmid (1980) indicates that there is evidence of an increasing
stabilization and decreasing rotation of the foreign labor force in
Switzerland. In 1970, 25 percent of the foreign-resident population
was 16 years of age and under. In 1960 the percentage in this age
category was 13 percent. The increase in the number of foreign workers with the rights of permanent residence has gone from 175,000 in
1970 to 371,000 in 1980, 403,000 in 1983, and 414,853 in 1984.

Another indication of stabilization is provided by the graphed
trends of foreign workers by category published by the Conference
Board (1980, p. 3). The graph indicates the changes from 1970
through 1980. Resident alien workers moved from 200,000 in 1971
to about 350,000 in 1980, but annual contract workers went from about
400,000 in 1971 to about 120,000 in 1980. Seasonal workers began at
about 125,000 in 1971 and ranged up and down during the decade, fin-
ishing at about 75,000 workers in 1980. The border-crossing workers

---

upgrading the few low-income residents. Unlike the United States and
the United Kingdom, Switzerland's acceptance of immigrants is primarily on the basis of assimilation rather than on multicultural acceptance. This is, of course, a result of pressures deriving from
the concerns for Überfremdung ("excess foreign influences") found in
a small but vocal sector of the society. This concern is also a product, in part, of the closeness of local community opinion, which will
probably change with time.

tended to follow the 100,000 mark from 1971 to 1980. Thus, we can see that annual workers have generally been converted to resident workers in the decade, and the only unemployed who may have been exported are in the lessened number of seasonal labor contracts and a small percentage of the annual contract workers. With further stabilization of foreign workers and their families with residence permits, it is only a matter of time before these workers and their Swiss-reared children will be integrated into the Swiss society.

The author interviewed a number of resident foreign workers and their families. A few were formerly from Italy, but they had been so long in Switzerland as to view themselves as Swiss-Italians. One family had been in Switzerland in the same town for 28 years. The man had gained advancement in the textile factory where he had moved from laborer to color technician. His wife worked in another factory in town as a packer of finished products, and in her spare time, she took on housecleaning jobs. The two daughters attended school and helped maintain the family apartment. The husband and wife spoke German and Italian, but the daughters spoke German, English, and some French. The two girls were preparing for secretarial and sales work, respectively, and had already chosen their apprenticeship assignments. The family talked of their vacations, which they had spent in the United States one summer, in Canada another summer, and in Italy in other years. The family was saving its money to buy a car and to help their daughters if and when they married.

Almost every week, in each of the hundreds of community newspapers in Switzerland, one can find announcements such as the following:

COMMUNITY HALL OF UNTERBERG

Invitation to the official election meeting of citizens of the city of Unterberg to be held Tuesday, June 28, . . . at 19:30 hours in the Community Meeting Hall.

Agenda.

1. Budget discussion authorization for the fiscal year of
   . . . beginning July 1, ... Copies of the proposed budget may be secured in advance at the administrative offices of the community at Community Hall.
2. Discussion and action on a city ordinance to permit the
   . . . , copies of which may be secured in the administrative offices at the Community Hall.
3. Discussion and action on the citizenship proposal of Mr. Walter Kurt Goldmann and his wife, Ingeborg Wohlberg Goldmann, and their children, Joseph age 5, Rosalind age 3, and Sven age 1. The family are all former citi-

zens of Hungary, and reside at their home on Brunnmatt-
strasse 23 in Unterberg.

Following item 3 would usually be three to five additional items of
proposed citizenship for people from West Germany, Czechosolvakia,
Italy, France, and other countries. After these items would be an
announcement that these citizenship-proposal files are available for
examination at the administrative offices of the Community Hall.

It is true that the position of the foreign worker in Switzerland
without permanent residency or without citizenship is anomolous, but
it is much less anomolous than the position of the undocumented worker
in the United States, who can never secure acceptance under present
law. It is no more anomolous than the position of the resident alien
in the United States who has never achieved citizenship. It is no
more anomolous than the position of the Swiss citizen in a foreign
country who must seek residency permission, even in the United
States. What the Swiss offer the French, Italian, West German, or
Austrian foreign worker is probably no more than those nations offer
the Swiss persons seeking permission for employment in those countries.

The Swiss probation period for foreign workers is probably a
necessary and practical means of securing the labor it requires for
what the nation considers to be an unusual and exceptional period of
economic expansion. The point is that the Swiss do not count on continued expansion or even maintenance of the economy at its present
level, and they seek to be secure in not having to again face the hunger and economic depression encountered earlier in its history when
their sons had to be sent away as mercenaries or economic refugees.
The constraints on immigration are to be expected in a rational and
planful society. Would anyone admire them more if they were to open
their doors only to end up in an egalitarian famine?

The Swiss school systems have developed special policies to
serve foreign children (Schmid 1980). Although these children attend
the same schools and follow the same curricula as the Swiss children,
many are provided with instruction in their mother language in addition
to their regular subjects. There is no formal discrimination of foreign children. Many of the foreign children have difficulty because of
the extra learning burdens of an additional language and perhaps because of the lesser pressure for academic achievement by foreign parents. Nevertheless, a study of Swiss and Italian children in Zurich
indicates that Italian males completed educational and career training
at a rate of 87 percent and Italian females, at a rate of 78 percent.
This rate is, of course, not as good as the Swiss male rate of 97 percent or of the Swiss female rate of 91 percent, but it is better than
U.S. high school and career-training rates. Los Angeles, for

example, reports a 60 percent dropout rate from high school and high levels of functional illiteracy among those who do complete high school. The Swiss failure to achieve equally positive training completion among Swiss girls is primarily owing to the marriage pattern among many Swiss girls, which induces them not to continue with education or training. With Italian girls, a particular cultural pressure is encountered from Italian parents who seek to keep their girls home, to prepare them for marriage, and to "keep them out of social circulation" as soon as they arrive at marriageable age (Gurny, Cassee, Hauser, and Meyer 1983).

Schmid (1980) reports that a high percentage of Italians, both at the moment of entry and approximately five years later, have no definite intended length of stay. There is generally a strong shift toward an increased time horizon the longer the actual length of stay. Thus, the period of contractual service for foreign workers is one where they have "one foot in Switzerland and another back home in their homeland" (p. 12). Only gradually do they begin to redefine the situation (and their intentions), just as was described by W. I. Thomas in his sociological analysis of the "definition of the situation," as it occurred in a Polish peasant study (Thomas and Znaniecki 1918-20).

In the 1970s immigration to Switzerland became an issue of national debate. In June 1970 a proposal to restrict the proportion of foreigners to 10 percent of the population was defeated by a majority. In 1972 another similar referendum was defeated by a margin of two to one. Another similar proposal in 1977 was defeated by a decisive 71 percent. Thus, it is obvious that Switzerland is slowly but surely moving to some action to further integrate the resident foreign workers who have made Switzerland their home.

Obviously, there are scattered pressures within the Swiss population to keep the foreign workers "in limbo" as aliens who remain available to do the unskilled work at lower wages and under less favorable working conditions. In fact, many of the Swiss do the less desirable work. But at the same time, there are forces in Switzerland whose sense of justice and honesty do not permit a continuance of second-class employment—and this body of opinion is obviously growing. Similarly, there are many Swiss who view the unwillingness of many Swiss children to begin their careers at the bottom rungs of the occupational ladder as inappropriate and "un-Swiss." Their view is that Swiss youth can (and do) do "dirty work" in their military training and annual military duty and that they can begin to do their share of these jobs. It is also obvious that Swiss factories and businesses seek a maximum of efficiency, and in the process, these businesses would not hesitate to upgrade a foreign-resident worker if there were no equally competent Swiss workers available for the position. Because the foreign workers are constantly screened by contract renewal,

it is obvious that their contributions to the Swiss businesses are becoming increasingly productive. It is obvious that, in time, the acceptance and assimilation of resident foreign workers will be achieved.
Walliman (1974) reports that total exclusion of foreign workers is not
a realistic expectation. What will happen to remove the cash barriers
and denominational and language idiosyncrasies insisted upon by many
Swiss communities for naturalization is yet to be determined, but
Hoffmann-Novotny and Killias (1979) believe that, in the long run,
political integration will become desirable for both the foreign-resident minority and the Swiss society.

Walliman (1984) believes that the resistance to full acceptance
of foreign-resident workers is because of an undercurrent of antagonism against them because of the structural tendency of industry to
use foreign workers to raise industrial productivity by lowering production cost. His conclusion is based on Marxian analysis, but it is
obvious that Swiss workers and labor unions will not permit development or expansion of labor-use trends that differentiate between Swiss
and foreign-resident workers. The obvious Swiss rejection of apart-
heid patterns in South Africa, where two levels of employment are
maintained and where one group of resident workers is disenfranchised,
will obviously be matched with a rejection of such patterns at home.
Kramer's (1980) observation about the Swiss, that they intend to remain honest at home even though they may make a profit on funds
less honestly secured by others in other lands, will probably be upheld in their treatment of foreign-resident workers. The Swiss know
that their banking business and their continued labor peace require
that they deal fairly with their fellow resident workers, who have
come to accept Switzerland as their home in all respects except one.
The control of external labor brought in to serve the transitory demands of industry during an uncertain semiprosperity of mixed economic reviews is yet another matter. Switzerland is wise to limit itself to temporary labor contracts with untried foreign workers rather
than to convert to a system of uncontrolled immigration. The latter
policy can only lead to problems of welfarized dependency for the foreign workers who fail to retain employment or for the Swiss workers
who, in a depression, may become displaced by new immigrants ready
to accept lower-status and lower-paid employment.

In any case, Switzerland is in a strong position to serve its current population (both native and foreign resident), whether in an economic expansion or a recession or depression. In the case of a recession or depression, it can probably better provide for its residents
(native and otherwise) than can the United Kingdom, France, the
Netherlands, or West Germany (countries that have kept less control
over their borders and that have relied on immigrant labor rather
than on contract workers).

The Swiss are wise to limit their seasonal and contract workers to numbers that are commensurate with current employment potentials. To do otherwise may be harmful, not only to the Swiss who enjoy full employment but also to their resident foreign workers who might otherwise become unemployed. To retain a two-level residency-citizenship policy is hardly democratic, but to shift to an open immigration policy, as was discussed by Pope John Paul in 1984 (see Schance 1984), would be equivalent to killing the goose that lays golden Swiss eggs.

REFERENCES

Conference Board. 1980. International Labor Migration, Economic Road Maps, nos. 1890 and 1891. New York: Conference Board.

Gurny, Ruth, Paul Cassee, Hans-Peter Hauser, and Andreas Meyer. 1983. Karrieren und Sackgassen: Wege ins Berufsleben jünger Schweizer und Italiener in der Stadt Zurich [Careers and blind alleys: leaving the professional life for younger Swiss and Italians in the City of Zurich]. Research Material 4 and Final Report of the Project, vol. 1963. Zurich: Soziologisches Institut der Universitat Zurich.

Hadorn, Urs. 1983. "Die Praxis zum neuen Asylgesetz–Versuch einer ersten Bilanz." Zeitschrift für öffentliche Fürsorge 80: 6–9.

Hoffmann-Novotny, Hans-Joachim, and Martin Killias. 1979. "Switzerland." In The Politics of Migration Policies, edited by Daniel Kubat, Ursala Mehrlander, and Ernst Lehinadier. New York: Center for Migration Studies.

Kramer, Jane. 1980. "A Reporter in Europe: Zurich." New Yorker, December 15, 1980, pp. 118–35.

Kümmerly & Frey. 1984. Switzerland 1984: People and State Economy Culture. Bern: Kümmerly & Frey.

Ricq, Charles. 1982a. "Frontier Workers in Europe." West European Politics 5: 98–108.

_____. 1982b. "Federalism and Trans-Frontier Communication." In Cooperation and Conflict in Border Areas, edited by R. Strassoldo and G. Dellizotti. Milan: Angeli.

Schance, Don A. 1984. "John Paul Chides Swiss for Complacency: Wealth, Neutrality, Treatment of Aliens." Los Angeles Times, June 17, pt. 1, p. 4.

Schmid, Carol. 1980. "Even in Paradise: The Immigrant Worker Problem in Switzerland." Paper presented to the Society for the Study of Social Problems, New York, August 24-27, 1980.

Thomas, W. I., and F. Znaniecki. 1918-20. The Polish Peasant in Europe and America. 3 vols. Boston: Badger.

Walliman, Isidor. 1974. "The Import of Foreign Workers in Switzerland: Labor-Power Reproduction Costs, Ethnic Antagonism, and the Integration of Foreign Workers in Swiss Society." In Research in Social Movements, Conflicts, and Change, edited by Lou Kriesberg, vol. 7. Greenwich, Conn.: JAI Press.

5

THE SWISS
LOCAL WELFARE
SYSTEMS

INTRODUCTION

Swiss public assistance at the local level is very much unlike the
public welfare programs of other industrialized nations. In the United
States, West Germany, the United Kingdom, and many other countries,
the public welfare program operates according to a complex set of
regulations emanating out of the national capital. This is not the case
in Switzerland, where policies and regulations are developed and administered locally and funded by local and cantonal taxes. Public aid
in other countries, including the United States, was at first delivered
as part of a social service process that sought to help the client to
again become self-sufficient. Unlike this pattern, which is still the
rule for Switzerland, public welfare systems in other countries have
been redesigned on the model of social insurance, where aid is distributed as if it were an earned benefit and a right, with little or no
differences in the manner or relative amounts given to families of the
same size: the check is mailed, and only at infrequent periodic re-
evaluations is there contact between the welfare system and most of
the clientele. In the Swiss system, this is not the case. The client
is periodically seen by a worker responsible to local authorities. The
client is held responsible for progress toward reestablishing on a
self-sufficient basis, and aid is related to the goals of rehabilitation
accepted by the worker in the relationship with the client. Swiss institutions, including family, schools, youth authorities, and voluntary
social service programs, constantly seek to prevent client dependency.
Despite all of this preventive effort, some Swiss do find themselves in poverty. Such persons are served not by the federal social-
insurance program, but by the local communities and the cantons.
Public assistance in Switzerland, unlike social insurance, is not a

concern or responsibility of the federal government. There is not even a federal data-collecting service relative to public assistance. Control of public assistance by the local community and the cantons has been jealously guarded ever since the creation of the state. Although all individuals have freedom of movement within the country (as in the United States and most of the Western world), the individual's home community or canton carries the bulk of the responsibility of providing for his or her care, either directly, if the individual is residing there, or on a reimbursement basis for the first five years if the residence is in another community. A person's established place of residency is that community and canton where he or she has been economically self-sufficient for a specified time or where he or she has been aided on a reimbursement basis by the home community for the specified time (Keller 1982). The determination for responsibility for reimbursements by the canton of residency is enforced by an intercantonal pact.

Because public assistance is funded from local and cantonal sources, it becomes a matter of concern for each local community to prevent public dependency, if possible, and to work with those who are dependent so that they can become self-sufficient as soon as possible. If a family is so beset with problems as to be unable to become self-sufficient in a short time, then the community authorities make special efforts to see to it that the children in that family are reared in a manner that will make them self-sufficient as they reach adulthood.

The Swiss public welfare programs of the local communities have a number of policies that they hold in common. The administration of these policies does vary from community to community, but the general effect is the same in terms of how the welfare workers view their assignments and how the clients view their grants.

These policies include the view that all public welfare aid is temporary and will continue only as long as the cause of impoverishment lasts. In the case of the elderly and some of the physically or mentally handicapped for whom eventual self-help would probably be ruled out entirely, public aid may last a lifetime. Although the social-welfare system may be viewed by some, especially in other nations, as an income-redistribution mechanism, this is not the case with Swiss public welfare. Thus, for the Swiss nonhandicapped and non-aged, any aid plan, once developed with the client, is quickly followed by a discussion of how long the aid is to be given and how soon the client will be able to again become self-supporting. This kind of a discussion involves questions of education and training completed, past employment and work skills achieved, further education and training required, and plans for finding employment again. If full-time employment is not immediately possible because of child-care arrangements, then school and day-care resources are provided in order to make part-time work possible. Even part-time employment is usually viewed as temporary, in that the client is expected eventually to become fully self-sufficient. There is essentially a contract between the client and the agency that covers the client's and the agency's responsibilities in regard to the amount, manner, and period of service to be given by the agency and in the client's movement toward independence.

The nature of the public aid given in Swiss programs also differs from that of many other Western countries. In Switzerland, it is incumbent upon the client to work toward rehabilitation and self-sufficiency. If he or she does not, the grant can be cut or eliminated. His or her children also must be reared to become self-sufficient. No such responsibility is placed on the welfare clients of other Western nations.

Swiss aid is described in most local community welfare ordinances as advice, counseling, information, and other social services, including if necessary, financial and material aid. There is usually no uniform schedule of grants, merely a set of guidelines that are considered along with the client's goal and plan. Considerations in calculating the amount of temporary aid include not only the general and special needs of the client, but also the amount of money he or she could eventually earn and whether a welfare grant larger than the beginning salary might discourage efforts toward training and employment.

The patterns of welfare administration in Switzerland are different from other welfare systems. In the Western world, public welfare is generally given on the basis of equality for each category of clients. The delivery of the welfare grant is seen as the primary purpose of the agency rather than as one of the tools of rehabilitation. Rehabilitation and social services in other countries are elective, if available, and the grant delivered to the client is related only to the specific plan and needs of that particular client. Finally, the welfare grant in other countries remains the same for as long as there are no reports of change in the client's situation. It can go on indefinitely (and it frequently does). This is not the case in Switzerland, where the amount of the grant is related to the client's progress.

Another major policy difference between the Swiss and other welfare systems relates to the responsibility of relatives and family. In some locations, this extends even to affluent brothers and sisters who are called upon to provide aid to needy siblings. Parents and grandparents are required to provide for their children and grandchildren, and in the case of grown children in need, parents are called upon to provide aid if they have the resources. Adult children are also required to aid their parents and grandparents if they are in need. Ex-husbands, in cases of divorce (and husbands and fathers in cases of separation) are required to supply adequate spousal and child support. In instances where child payments are not forthcoming on a regular basis, the public welfare agency is required by law to deliver the court-specified sums to the mother and then press for repayment. Unlike the U.S. experience, these payments are usually effectively recovered for a number of reasons, which will be discussed later.

The enforcement of the means test for public welfare in Switzerland is easily carried out. This is partly aided by the tax collection system, which is located in the individual communities. Each year every person files an income and wealth tax report. Local assessments are first levied on this basis, then cantonal assessments occur, and finally the federal assessments are made. For the sum of Fr5 (exempted in the case of welfare authorities), a copy of any individual's tax reports can be secured in most cantons by anyone without having to state a reason. If a report contains inaccuracies, these can be challenged by any other taxpayer. Thus, the tax record is an unusually authentic source for information on a person's resources and economic ability to support himself or his family.

Public aid plus interest in Switzerland (unlike other Western countries) can be recovered by the authorities from an adult client or a direct relative if he or she later becomes or is proved to be affluent. Thus, some clients tend to ask for less aid than they would if they knew that it did not have to be reimbursed. Similarly, public-aid funds that are secured under false pretenses or data can be recovered with interest by the local community. This keeps fraud down to a minimum.

Public welfare is conducted on a confidential basis in Switzerland as it is in most other Western nations. It is illegal under most community ordinances for a public welfare worker to release information without client permission, even if such information is requested by the local public welfare board members. Neither may a public welfare worker interfere with the constitutional or personal rights of any client. This is similar to the public welfare policies of other Western nations. The difference in the Swiss case, however, is that the client is required to cooperate with the welfare worker not only in regard to determination of needs and resources, but also in relation to improving his or her own situation and that of his or her family. Thus, a client who does not cooperate in working toward self-sufficiency may find that the nature and extent of the welfare grant has been changed, or that the welfare grant is to be dispersed in small increments by an appointed guardian, or that the welfare grant is now to be in the form of materials and vouchers. His or her only recourse is to appeal beyond the worker or supervisor to the public welfare commission or town council. Under such circumstances, the matter of confidentiality becomes moot.

Similarly, when necessary, the welfare worker is authorized to release information to youth and community authorities without permission of the client when the client has not been acting responsibly and when guardianship is to be considered. Here, too, the issue of confidentiality becomes moot.

From the above discussion, it becomes clear that Swiss public-welfare workers and administrators exercise considerable discretion in their relationship with clients. This is quite unlike the client-worker relationship in most other Western nations, including the United States, where the interaction is routinized and focused primarily on regulations of eligibility for aid and the authorized levels of grants. (In the United States, social services are now separated by law from eligibility processes, and social services are rendered to clients only on an elective basis.) In the Swiss public agency, almost no aspect of the client's life or the life of his or her family is prohibited for discussion by the worker. Unlike the restrictions on public welfare workers in other countries, in Switzerland the worker has generally unlimited controls in requiring interviews with the client and collateral visits with relatives, employers, teachers of the client's children, and others who may in any way affect the client's progress toward again becoming independent. Thus, the welfare worker, both formally and indirectly, can shape the way in which public aid and public social service are used by the client and his or her family. If a client seeks to go beyond the welfare worker's constraints or even beyond the agency's constraints, he or she then comes face-to-face with community authorities who are almost always in agreement with the goals of client self-reliance. In the author's interviews with public officials, it was learned that they are even more imbued with the need to reduce client dependency than are the welfare workers themselves.

The matter of confidentiality is particularly important in Switzerland because of the issue of stigma. Welfare in most Western countries is somewhat stigmatized by the general public, and this stigma serves to hold down the amount and degree of welfare sought by middle-class persons in need. This stigma has the opposite effect on the chronically welfare-dependent, in that it ensures their cultural separation from the world of work and upward mobility. In Switzerland, where most of the population is middle class and where public welfare dependency has a general air of stigma, there is considerable danger that becoming known as a welfare client may negatively label a person among friends, neighbors, community, and potential employers. Thus, in Switzerland stigma serves to promote self-sufficiency and client cooperation toward that end.

The view of Swiss public welfare administrators toward their clients seems quite different from that of most public welfare officials in other Western countries. On the causes of poverty, most Western public welfare administrative writings reflect the circumstances and conditions of poverty. It is a view that focuses on the structure of society, factors of unemployment, the lack of attractive low-level employment, the failings of public education, the findings of inadequate vocational training opportunities, and the lack of adequate housing opportunities for the welfare clientele. Swiss public welfare writings and discussions reflect the view expressed by Heinz Strang (1970) that poverty often is caused by a variety of factors, including structural unemployment, individual problems and conditions that prevent adequately remunerative employment, and dependency that may be interactive with the availability of public aid. Swiss welfare personnel, unlike their counterparts in other developed countries, tend to individualize interpretations of why a particular client is in need. The major difference between the two views is that Western administrators, other than those in Switzerland, tend to accept welfare need as a given and to accept the view that in any modern society that is beset with some dysfunctionalities there will always be a sizable proportion of the population in need. Swiss administrators, on the other hand, tend to view their world as a place where individuals have a choice; and they feel that if some are in need, it is because they have made a wrong choice or because they have been unfortunate in their circumstances or because of a combination of these factors. Keller (1982), for example, states that the highest principle of social workers is to help people to help themselves. Rather than guarantee the client a minimum income, Keller emphasizes the need to guarantee that the client will be led toward a meaningful life. Public charity, he indicates, should be carefully given with the individual carefully taken into account, so that he or she can be speeded toward resumption of social and economic independence.

What is to be done for and with the poor? Western administrators (aside from those in Switzerland) see the solution as a simple matter. Because they define poverty as a lack of money, the way to solve it is to redistribute some of the society's funds to the poor. The Swiss welfare administrators define poverty as a condition of multiple causation and seek to solve the problem by determining the individual causes in each case and to work accordingly. In a sense, the non-Swiss welfare world deals with poverty as a fated condition, and the Swiss welfare world views poverty more rationally, as if it were some societal and/or personal social or psychological dysfunction needing to be dealt with according to its cause. Other Western administrators view their clients as receivers of benefits rather than as actors in their own life choices. Swiss administrators view their clients as differentially able to make changes in their life, space, and circumstances, once they have been encouraged and helped to determine what has to be done.

Western administrators tend to emphasize the goals of efficiency in the delivery of benefits, more egalitarian handling of clients in a massive system, and improved management of their assignment in the fulfillment of diverse regulations. Under the Swiss system the emphasis is placed on individual adjustment to the society so that the client can learn best how to live productively.

The welfare process is seen as a necessary function in the Western world, aside from Switzerland, and welfarization is viewed as inevitable in an imperfect world. In Switzerland welfarization (the process of persons becoming used to welfare dependency) is viewed as something to be avoided.

In fact, many Swiss welfare workers view welfarization as an iatrogenic disease (a disease caused by professional treatment). In an interview with the author, one Swiss welfare official drew a parallel between welfarization and excessive bedrest after surgery: "If the client gets used to inaction he will lose the ability to fend for himself."

In addition to the communal and cantonal public welfare agencies, there is also a diversity of voluntary agencies. These agencies provide a major part of service to the population of Switzerland in both the prevention and treatment of social problems. There are a diversity of institutions, not only in Zurich but also in the towns and regional centers. Many of these agencies operate under enabling laws and civil statutes. There is a great diversity of services, some with narrowly defined purposes such as operating a local old-age home or drug-treatment clinic, some with broad regional programs such as the French-speaking trust for social and moral hygiene or the national organizations with broad scope and with local service programs. Examples of this latter category are Pro-Infirmis (aid to the disabled), Pro-Senectute (aid to the aged), and Pro-Juventute (juvenile assistance). Pro-Juventute and Pro-Senectute are legal trusts operating through largely autonomous suborganizations in the cantons and districts. Pro-Infirmis has decentralized advisory offices throughout the country, which are funded nationally and are not dependent on cantonal or local support (Büchi et al. 1979). Many of these organizations are an amalgamation of loose, large organizations that are themselves "parents" to other agencies. For example, Pro-Infirmis is not only a union of social services for the disabled, it is also a parent organization of 13 other organizations, some with their own local service branches specializing in particular forms of disablement. For some of these associations, the central office of Pro-Infirmis is their central office; for others, their central office is located elsewhere. Many of the organizations provide separate regional offices in the French areas of Switzerland. There is no prototype. Each association has developed according to its own constitution and by-laws.

Pro-Juventute receives its money from a number of sources. It operates an annual sale of stamps in cooperation with the Swiss postal and telegraph services. In 1981/82 this brought in over eleven million francs. Private legacies produced Fr4,654,000 in that year. It received subsidies of over seven million francs, and other gifts and income brought in over five and one-half million francs, making a total income of over twenty-eight million francs. Its funds were mainly spent on national, cantonal, and regional activities among youth and in direct welfare grants to needy youth (Statistisches Jahrbuch der Schweiz 1982).

Pro-Senectute had a 1981 income of over fifty-seven million francs, which was received partially from contributions, about thirty-five million francs and eight and one-half million francs from canton and community subventions, respectively, and almost fourteen and one-half million francs from confederation subsidies.

Pro-Infirmis received a total of over Fr23 million in 1981, primarily from private and public contributions, which it spent as follows: subsidies for programs for invalids, Fr1.3 million; grants to handicapped individuals, Fr6.6 million; and welfare grants to handicapped persons, Fr3.1 million.

Each of the three agencies provides directory service for persons in need of special help.

The process of finding resources to serve individualized problems is made easier in the Zurich canton by the availability of the Handbook of Social Policy, published by the Information Office of Social Policy. This handbook lists all of the social service organizations in Zurich canton by types of service and by locations. The listings total approximately 500 agencies. Unfortunately, there is no similar handbook for the other cantons. The Institute for Family and Marriage of Zurich also publishes an all-Switzerland directory of family advice and counseling bureaus, in which family counseling agencies are listed for 20 of the 26 cantons. In many of these cantons, there are numerous listings for each of the towns with sizable populations.

Many of the voluntary agencies (according to Büchi et al. 1979) have become involved in issues relating to the needs of special client groups to the degree that they have succeeded in securing expansion of the local welfare programs to serve these special groups.

The public social services are generally imbued with the reputation of being "official" rather than informal because of the fact that many of them (the communal offices, youth-serving agencies, and so forth) are empowered to take measures of social control. Many are even obliged to do so as part of their mandate. The public welfare agencies, often being the service of last resort, are frequently concerned with the more complicated, multiproblem, cost-intensive cases. Unlike the public social programs in other countries, the constraints and restrictions on the work of the Swiss agencies are not bound up in legal constraints. Büchi et al. report a large scope of discretionary action open to public-agency work. Also typical of public-agency work in Switzerland is the considerable control and influence in the hands of directly (or indirectly) elected supervisory bodies (commissions, authorities, inspectors, and so forth). Their influence is even greater in the rural areas of Switzerland, where professional social service personnel are less frequently found.

The voluntary agencies in Switzerland are more numerous and active than in other countries where welfare-state systems have taken over much of the responsibilities and authority. The nature and emphasis of the programs differ in each of the cantons and communities, thus reflecting the different interests and concerns of the regions. In Switzerland the voluntary agencies were almost exclusively dependent upon private sources of income (members' contributions, bequests, donations, and firm assets) until World War II. Since then, an explosion in costs of programs has occurred, and the added costs have been largely absorbed by public fund subventions. In other countries the expansion of costs and programs developed primarily in the public agencies, and private agencies have remained relatively small; but in Switzerland the private agencies have continued to be predominant. The subsidies generally are supplied by the communities, the cantons, and in some cases, from federal sources. Federal insurance programs are particularly involved in the reintegration of disabled persons. The necessity of dealing with a variety of "backers" may involve administrative burdens and require involvement of a variety of community or regional interests in program planning, but it also ensures the agencies a degree of independence as well as a considerable degree of community responsibility.

Thus, the improvement of private-agency programs is dependent upon public understanding and acceptance rather than upon "behind-doors" budgetary negotiations and legislative trade-offs, as they occur in public programs in the welfare states.

The funding of public agencies in Switzerland derives from the respective communities and cantons. Federal support occurs only in the case of refugee services and care of Swiss returning from abroad who are without funds. In the case of foreigners, the costs of their care falls primarily on the communities and cantons, except where there are legal agreements in effect with neighboring countries for the funding of cross-border services. Many cantons have laws requiring restitutionary funding by responsible relatives of welfare clients; and according to Büchi et al., this is usually carried out both selectively and discreetly, based on the degree to which the relatives have adequate resources. Public funding of voluntary agencies is generally dependent on evidence that the service is needed and that it is not already being provided by some other agency. This causes agencies, both private and public, to consult with one another and to limit their spheres of activity. In some instances public authorities have asked specific private agencies to install and operate programs when they have been deemed necessary. This has frequently occurred in the matter of establishment of youth and family centers, requests for which have been directed to Pro-Juventute, known for its pioneer work in this regard, and in the matter of management of domestic services for the aged, which have been directed to Pro-Senectute. In these instances, public subventions have been provided.

A new development has occurred in the 1970s. Disadvantaged groups that do not believe that their interests have been sufficiently provided for have formed self-help organizations. Some of these groups have existed since the 1920s in Switzerland, but prior to the 1970s, they were limited and often dependent upon the social service agencies for leadership.

In the 1970s many of these groups became quite active, and new groups formed as well. Unlike the previous pattern of activity, however, the new programs tried to focus on areas of disagreement with the agencies and to serve as organizers for the client groups seeking social actions, service additions, mutual aid, and improved benefits. These groups include associations of single parents, discharged prisoners, and others. These groups seek attention and improved funding from public authorities. This development leads to some competition for funding, creating in effect an economic marketplace of social needs. This kind of development does much to offset the bureaucratic format of operation so prevalent in the welfare programs of other nations.

The diversity of services and funding sources, as well as the development of self-help groups, does make for a lack of coordination and impedes the development of social services planning, but there are advantages in the diversity. Competition for private and public attention and funding means that large numbers of trustees, donors, commissions, advisory boards, and public officials are involved and concerned with the quality, nature, extent, and scope of the agencies they fund or sponsor. In a sense, this creates a much more democratic pattern of services, more responsive to client needs and public interest than is provided by the relatively rigid public structure of the welfare state, where the public is involved only indirectly through their ballot box.

Coordination at the local level occurs through the public welfare commissions, which both fund the private agencies and hold responsibility for supervision of the public welfare agencies. In addition, there is an informal level of coordination that occurs in the many instances when the formal and informal leaders and members of the community meet and participate in social and communal affairs. The cohesive effects of citizen interaction when community men and women get together under the aegis of military activities, civil defense programs, volunteer fire departments, guild activities, and civic club meetings has been noted. Many of the same people who are involved in social services are also involved in these community affairs. In addition, there are national and regional coordinative bodies. The Swiss Conference of Public Welfare and the Conference of Cantonal Welfare Directors serve to coordinate public welfare services. The Swiss National Conference on social matters serves to bring together public and private agency leadership.

Despite the fact that there is no uniformity of service patterns (that is, communal care of the sick might be under a public agency in one locality and a private agency in another), there is an evident emphasis on the part of local public and private social service leaders to ensure that basic services are available in each locality. Some localities have entered into the pioneer work of creating regional associations to provide services where the communities are too small to provide them individually. The Oberhasli district welfare program in Bernese Highland is one example of cooperative programs. Another is the drop-in center for drug counseling organized in the Aarau area. Büchi et al. report that the Vaud Canton is engaged in integrating its public and private social services by district and that one such integrated center in Yverdon was already established in 1979. Some of the integrated services have been promoted by federal assistance under the program to assist mountain areas.

An interesting result of this diversity of services and sponsorship of programs, according to Büchi et al., is that the agencies are now more involved in working with the clientele rather than for them or on them. It is obvious that interactive involvement with the clientele and community citizens is more likely to be more effective than the bureaucratic formalized processing that occurs in highly centralized bureaucratic service institutions.

The diverse mixture of private- and public-agency sponsorship and funding presents the Swiss programs with questions on the definitions of service goals. In other nations with highly centralized and formal public programs, the roles of the agency, the client, and the community are clearly defined. The agency provides specific services (whether in kind or cash), and only those services in a manner and format outlined under federal and state regulations. The client is allowed to apply, and the application is processed. The client receives only those items and services requested that are provided for under the regulations. His or her influence on the agency and vice versa are both minimal. The community is not involved at all. If someone were to express an interest in a client or in the program, that person would probably be politely (or not politely) rejected. The voluntary agencies, whether centrally directed or local in nature, also provide special services, and these are discretely defined and not usually involved with the public social services. It is almost as if there were separate worlds, public and private, and the client visits both but is seldom a part of either.

In many instances the public agencies have a social-control function, and the authority of law is vested in them. These include such programs as child-protection agencies, public adoption agencies, child-support programs in divorce courts, probation and parole agencies, and so forth. These agencies are, in the eyes of the client, imbued with coercive authority and directives, and it is difficult for the agency to help the client understand that they exist to help the client resolve problems (whatever the agency assignment may be). In the case of public welfare in most Western nations, the directive to the public agency is to process the aid grant, and social services are purely voluntary, subject to the client's acceptance. Thus, in the public welfare field, there is no supervisory service provided and required, and clients are left to themselves to choose to move toward rehabilitation or not, as they wish. Even the social services that are provided are concentrated not as much on rehabilitation as on help in adjustment to life on public welfare. Because of tight regulations and large caseloads, what services are given are on a mechanistic basis and seldom, if ever, provide for a continuing client-worker relationship.

The nature of the Swiss client-worker relationship is far different. Whether in a private or public setting, the focus of the agency's service is primarily to use the client-agency relationship toward the goal of client rehabilitation. As such, it makes of the worker (and agency) a type of parent figure, in that the agency provides the resources for rehabilitation and is answerable for their use. The client, in turn, is subject to the agency's approval and guidance in his or her movement toward a return to self-sufficiency and is motivated to act in rehabilitative ways if he or she is to continue to receive agency resources. Thus, unlike the relationship in other Western nations, in Switzerland the client is answerable for what he or she does with self, family, and agency resources received. The agency worker is answerable to a supervisor for what is done in work with the clients, and the supervisor is answerable to the agency director for what the workers are accomplishing in client rehabilitation. The director, in turn, is responsible to the board or commissioners, and they are responsible to the citizens of the locality. They can be depended upon to ask for answers to questions about expenditures and rehabilitative achievements, especially when the local budgets are under consideration.

In the Swiss public welfare agency, the client knows his or her worker and the worker knows his or her client. This is very much un-like the condition in other Western nations, where the client is likely to see a different worker at every agency contact. The chronically de-pendent client is often one whose level of socialization into the main-stream world is limited, who has difficulty relating to an institution. He or she can, however, relate to a person. A person cannot be ex-pected to move toward self-dependence unless he or she knows (1) that opportunities for self-dependence exist, (2) that someone with authority is backing the move toward self-dependence, (3) that he or she can expect to be watched by that someone while dependent on wel-fare, (4) that certain discomforts and sanctions can be expected if he or she does not move toward self-dependency, and (5) that he or she can look forward to tangible rewards as he makes progress toward self-sufficiency. In the Swiss welfare agency, this kind of situation is fairly certain. In other Western nations, this does not occur. Be-cause of huge caseloads, impersonal institutions, frequent staff turn-over, huge bureaucratic "paper" and regulation requirements, and a sense of worker estrangement and alienation (described by Street, Martin, and Gordon 1979), the client is usually lost in the machine. The client can expect to be lost in the machine, and thus, there is no pressure for client rehabilitation under these conditions. In addition, the regulations setting forth the "right to welfare" in other nations make rehabilitation a taboo subject. The Western world provides welfare clients with workers who are uncommitted to rehabilitation goals and who retain low expectations for their clients. Only in Swit-zerland can the client expect to be held responsible for rehabilitation progress; only in Switzerland can the worker expect to be held re-sponsible by his or her supervisor and the local welfare-board mem-bers for supervision of client rehabilitation; and only in Switzerland has public social work (and the influential social work profession) held itself to the original social work goals of client rehabilitation. In many other countries, social work has become dichotomized, with "professional" social work's understanding acceptance of the public welfare client's rationales for not moving toward self-responsibility (a form of folie à deux) and with public welfare workers distancing themselves from their welfare clientele.

Thus, the structure of Swiss public welfare ensures that client rehabilitative goals will remain in the forefront of client-worker re-lationships. The role of the worker is necessarily dichotomous—as an agent of the community and as a helper to the client. This is a much more difficult relationship to establish and maintain than the simplistic welfare functions in the other nations, where separation of eligibility and services has occurred. In these settings the worker is either a social-control agent (as eligibility worker) or a "helper" (as provider of social services), but not both. For client change to occur, especially where the client is not self-motivated, the worker requires authority as well as skill. In the Swiss social service setting, both elements (authority figure and helper) are available in both the public agency and in the private agency, where public funds are utilized and where a close relationship with public agencies is maintained.

In this respect, the private and public agencies in Switzerland provide the gemeinschaft-type services so urgently required in the industrialized, impersonal, gesellschaft setting of modern life. Unlike the agencies in other Western nations, which have sought to convert to the impersonal mechanisms of the gesellschaft, the Swiss agencies have retained their gemeinschaft practices and community ties. The Swiss agencies have held to a pattern that helps them retain a measure of effectiveness that agencies in other countries have generally lost.

This combination of helper and community agent serving a gemeinschaft program makes it more likely that clients will be provided for when need is present, even if the client is reluctant to become a welfare dependent. In inquiries about client abstention from supplementary social security (Ergänzungsleistungen), numerous instances were found where local community health nurses, volunteers of Pro-Senectute and Pro-Infirmis, and tax agents of local communities have encouraged people to apply if they were aged, survivors, or disabled and without a social security prepayment history. The extent of such abstention from this program is believed to be very low (Charles 1981).

The public expenditures for social services in Switzerland are listed in the 1985 International Labour Office reports on The Cost of Social Security: Basic Tables (pp. 140–41). These are presented in Tables 5.1 and 5.2.

It is interesting to examine the costs of Swiss welfare in comparison with the welfare costs of a comparable population. Switzerland spent $1,003,200 in 1980. Its total population is listed as 6,460,000 in The World Almanac and Book of Facts, 1985 (Newspaper Enterprise Association 1984). This amounts to about 16¢ per capita. Los Angeles County, with a population of 7,477,421 spends $91,657,222 for Aid to Families with Dependent Children (AFDC), $5,952,633 for general relief, and $9,358,349 for adult supplemental aid (Los Angeles County, Department of Public Social Services 1984), making a total of about $107 million. This amounts to about $14.31 per capita.

The report on public welfare expenses indicates a rather steady pattern, with slight fluctuations, from 1979 through 1980. In U.S. funds, the public welfare expenses are estimated at $950 million. These expenditures do not, however, include the cost of operation of

TABLE 5.1

Swiss Public Assistance Programs: Income, 1980

(in millions of Swiss francs)

Income	From Confederation Participation	Participation of Other Public Authorities	Transfers from Other Schemes	Other Receipts	Total Receipts, 1980
Complementary old-age, invalidity, and survivors' assistance	—	—	10.3	—	10.3
Cantonal public assistance	3.7	654.2	—	167.7	825.6
Communal public assistance	—	988.9	—	—	988.9
Total	3.7[a]	1,643.1[b]	10.3[c]	167.7[d]	1,824.8[e]

[a]Prior figures were 6.9 for 1977, 11.0 for 1978, and 7.6 for 1979.
[b]Prior figures were 1,503.8 for 1977, 1,497.4 for 1978, and 1,523.7 for 1979.
[c]Prior figures were 17.3 for 1977, 17.5 for 1978, and 10.3 for 1979.
[d]Prior figures were 260.5 for 1977, 268.6 for 1978, and 232.2 for 1979.
[e]Prior figures were 1,778.5 for 1977, 1,794.5 for 1978, and 1,773.8 for 1979.

Source: International Labour Office, The Cost of Social Security: Basic Tables (Geneva: International Labour Office, 1985), pp. 140–41.

TABLE 5.2

Swiss Public Assistance Programs: Expenses, 1980
(in millions of Swiss francs)

Expenses	Cash Benefits, 1980	Total Expenses, 1980
Complementary, old-age, invalidity, and survivors' assistance	10.3	10.3
Cantonal public assistance	—	825.6
Communal public assistance	—	988.9
Total	10.3	1,824.8*

*Expenses for 1977 were Fr1,788.5; for 1978, Fr1,794.5; and for 1979, Fr1,773.8.

Note: In 1980 the Swiss franc was equal to ¢55.0 in U.S. dollars. Thus, total expenses can be estimated at $1,003,200.

Source: International Labour Office, The Cost of Social Security: Basic Tables (Geneva: International Labour Office, 1985), pp. 140–41.

the youth service bureaus, youth centers, and other direct service programs maintained by the cantons and communities.

ZURICH

Zurich, the largest urban unit in Switzerland, provides a picture of Swiss urban poverty and a public welfare system that yields a basis of comparison with poverty and welfare in the urban Western world. The research report of Hauser et al. (1980) provides a detailed study of poverty in Zurich. In this study, summary protocols of families and individuals in poverty were studied out of a total of 3,646 persons who were aided by the city of Zurich in 1976 and of 349 welfare clients registered in the 21 outlying communities of Zurich in the Zurich Canton. Thus, in the Zurich Canton there was a total of approximately four thousand persons who were considered to be in poverty in 1976. The total sample of client protocols examined came to 601.

Based on the research study data, in the Zurich Canton there were 1,106,518 people in 1976. This yields a 0.36 percent welfare dependent rate (less than 1 percent). To compare the welfare rate of the largest city in Switzerland with the second largest urban area in the United States, it is necessary to examine the following data:

Total population of Los Angeles County (as revealed by The World Almanac and Book of Facts, 1985) equaled 7,477,421.

Welfare statistics for Los Angeles County (Los Angeles County, Department of Public Social Services 1984, p. 1) were

AFDC	551,243
General relief	39,986
Adult supplementary	41,159
Total persons served*	632,388

The number of persons dependent on welfare in Los Angeles County (not counting the items listed above) comes to approximately 8 percent. Thus, the welfarization rate of Los Angeles County is over twenty-two times that of the canton of Zurich.

The Zurich cantonal study indicated that welfare families existed not only in the city of Zurich, but also in the area around Zurich. In

---

*Omitting food stamp recipients and a variety of special services, such as public housing, housing vouchers, winterization aid, school lunches, nutrition programs, and so forth.

the city, with a population of about 383,000 people, 3,646 persons were aided (about 0.95 percent). In the industrial cities, there was a population of 110,285 people, with 183 families under public aid (about 0.17 percent): in the residential communities, there was a population of 50,317 people, with 147 persons aided (about 0.3 percent); and in the rural communities, the total population was 11,226 people, with 19 persons aided (about 0.17 percent). From this it is apparent that the welfare rate approaches 1 percent in the city of Zurich but falls to less than one-third of a percent in the residential communities and less than one-fifth of a percent in the industrial and rural communities. In any comparison, the Swiss urban welfare rate is far smaller than that of the U.S. urban areas (if Los Angeles can be considered a representative sample).

The makeup of the welfare population in Zurich Canton is of interest. In Zurich City and rural communities, males predominate, and their cause of poverty is most often temporary unemployment. In the residential and industrial communities, females with children predominate, and the primary cause listed was delayed or nonexistent alimony payments. This problem will be discussed later. It should be noted that a cantonal ordinance in 1977 granted the welfare departments the right to advance up to Fr 500 per month per child where alimony from the child's parent is not forthcoming, but this law also gives the welfare departments the right to sue the parent. Apparently, this method of dealing with unpaid alimony payments is expected to resolve much of this problem.

The client age distribution was examined by the study. The researchers found that young adults in poverty (ages 20 to 30) were highly overrepresented in the rural communities, and that adults (ages 30 to 40) predominate in the poverty of industrial towns and Zurich City. People 40 to 50 years in age were heavily overrepresented on the welfare rolls in Zurich City and only lightly represented in the industrial towns. The researchers found that welfare recipients in the industrial communities are the least educated group, but in Zurich City unskilled persons represent only a small share of the cases (33 percent).

The reasons for welfare need included insufficient alimony payments twice as often in the industrial towns as in Zurich City. In Zurich City every other welfare recipient was unemployed. In Zurich City more than a third of the welfare recipients reported illness as a cause of their being in need. All communities in Zurich Canton had an estimated fifth of the welfare population with psychic illnesses. In the rural communities in Zurich Canton, 25 percent of all cases had a diagnosable case of alcoholism, and alcoholism was an important welfare reason in all communities. Drug addiction occupied only a relatively small share of the welfare cases in all Zurich Canton.

Few of the welfare population had criminal records. A surprising
finding was the small number of children in the welfare families. At
least half of the welfare clients had no children. A third had children
(only one or two children). Persons in auxiliary occupations (temporary help in restaurants, newspaper vendors, and unskilled labor-
ers) were overrepresented in the welfare clients (49 percent), despite
the fact that they were only 18.6 percent of the total Zurich population.
About 83 percent of all clients had received welfare grants in the past.
Of the total number of welfare clients, 18 percent suffered from physical or mental illnesses, and sickness and accidents figured in 53
percent of the clients studied. Apparently, a major function of welfare is that of providing medical care for those not covered by medical and accident insurance. In the few instances where clients had a
"good" education, the researchers found an accompanying alcoholism
or drug addiction. Apparently, well-educated and trained people do
not have to depend on welfare, except when unable to retain employment owing to substance abuse. The residential communities showed
the highest rate of clients with prior dependency on welfare. In other
communities of Zurich Canton where welfare policy restricted grants
to temporary need, there were fewer cases of past dependency. The
researchers wondered whether or not the residential communities had
welfare personnel who were less able to achieve rehabilitation with
their clientele. An interesting finding was that women claimed welfare benefits for economic and health reasons more often than men.
Men were apparently drawn to welfare need after socially deviant behavior, particularly in cases of alcoholism. The researchers indicated
also that unmarried persons constituted the most important welfare
groups—in particular, single men and divorced women. Unlike the
patterns found in the United States and other Western nations, these
researchers found few single mothers in the welfare population. The
researchers believed that such women were few in number in the population, and those who were rearing their children were also either employed or being helped by their parents. Divorced women (ages 20 to
65) with children were the most frequently occurring welfare clients.
They were usually on welfare for several years, primarily as a result of insufficient alimony. The researchers concluded that this problem among welfare clients would be eased with the new legislation
that will enable welfare advances to mothers with legal reimbursement
actions on such fathers.

The policies of the Welfare Agency of the city of Zurich, as published by that office (Urner 1979), are generally reflective of the welfare policies of other Swiss communities and of those set out by the
Conference of Swiss Public Welfare.

These policies interpret the various regulations on welfare in-
corporating the changes of the government council law of June 13,

1979, and the original Zurich poverty law of October 23, 1927. These regulations state that public welfare is an administrative matter of local government. Assistance is to be given on an individual basis, according to the peculiarities and needs in each individual case. The regulations require the recognition of the causes of welfare need, such as unfavorable education, inappropriate occupational choice, mental difficulties, physical maladies, addiction, mental illness, psychopathology, and so forth. The worker is required to avoid the tendency to judge the client as having maladaptive behavior and, rather, to give purposeful and consistent assistance (for the purpose of not only alleviating poverty but also rehabilitating the client to prevent continued poverty). Thus, the worker is directed to give the client personal assistance and economic assistance where needed. The statement emphasizes that the client's abilities, self-reliance, and positive faculties can only be strengthened with the client's cooperation. Quick action is encouraged. Prophylactic help to prevent a threatening emergency is encouraged. Help should include social counseling; supervision of wage and pension payments; debt amortization; rehabilitation; budgetary counseling; domestic guidance; and arranging for medical aid, legal counseling, job procurement, and so forth. In this area of services, the principle of voluntary participation of the applicant applies.

The issue of self-responsibility is ensured with the statement that "economic assistance is granted only if a person is unable to support himself or his dependents adequately or timely from his own resources" (Urner 1979, p. 243, emphasis added). Economic assistance means granting a minimum social existence, including expenses for living costs, individual needs, medical treatment, as well as education for children. "Financial help is to be correlated with the efforts of the applicant, particularly his duty to inform the authorities and compliance with the instructions of the welfare authorities. If necessary, guardianship (of the client), can be applied for" (p. 2., emphasis added). The forms of economic assistance are cash and, in isolated cases, foster homes, institutions, hospitals, and vouchers (if necessary) to safeguard proper usage of living cost assistance. The law of Zurich Canton states that "the duty of the community authorities is to render personal assistance . . . in consultation with the applicant without a specialized procedure" (p. 1).

The principle of subsidiarity shall be maintained—namely, that the family shall be self-supporting. Only when it is unable to do so can the welfare bureau funds be used. Also, the principle of complementarity applies—namely, that welfare funds are to be used to complement the client's available resources and income, rather than as a substitute for use of income or earning power.

Public economic assistance shall be used only after self-help earnings and collection of claims on third parties, including insurance claims, social security claims, wage claims, and so forth. These claims are to be ceded to the welfare authority that grants benefits. The applicant has to reimburse the welfare authority or offer assets for attachment if such assets are available.

The welfare regulations support civil law obligations between married couples, parents and minors, blood relatives in ascending or descending line, and brothers and sisters in favorable circumstances. While the welfare authority has the duty to secure the existence of the applicant, it also has to clarify the rights of the applicant toward others.

Decisions of the welfare authorities can be appealed to the district or government council. The welfare authority shall safeguard the interests of the applicant, particularly in cooperation with other social institutions. The Zurich public welfare authority and responsibility is presented in Figure 5.1.

The principles of public welfare law in Zurich and elsewhere in Switzerland differ considerably from the welfare provisions and supplementary benefits of the welfare state nations, as well as the U.S. public welfare programs. Primary in the differences is that of local regulations, budget, autonomy, and administration. The local agency and the client's treatment are not circumscribed by highly centralized regulations formulating client rights and agency formulas for service, as they are in other nations. Recognition of the cause of need is primary in the relationship between client and agency. This is not the case in other Western nations where the client has an entitlement to aid, and where the cause of need is not at all examined, as if the client had no basic responsibility toward self-sufficiency. Thus, the principle of subsidiarity is basic to Swiss welfare.

Another difference is found in the assignment of purposefully selecting a goal of rehabilitation of the client and using the assistance in fulfillment of that goal. This is not the pattern in other Western nations where rehabilitation and supervised use of economic aid is not entered into with the clients. The focus of Swiss welfare is primarily on rehabilitation, while the focus of other Western welfare is limited to eligibility and grant disbursement only. The use that a client makes of financial assistance in other Western nations may not be examined, but in Switzerland it is an important function of the social service process. Considerable discretion is granted the Swiss welfare service in its dealings with the client, but this is not the case in other Western nations. In Switzerland the primary focus is to help the client become self-sufficient again—and financial aid is merely one tool in the treatment of the client. In other Western nations, the primary focus is mechanical processing for eligibility and disbursements.

FIGURE 5.1

Social and Individual Help of the Welfare Agency of the City of Zurich

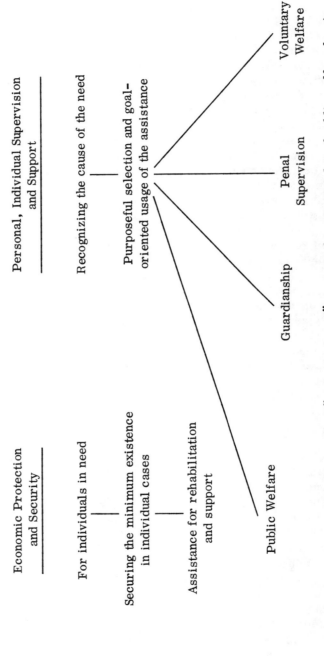

Source: Paul Urner, Grundsatze des Züricherischen Fürsorgerechts [Principles of public welfare law in Zurich] (Zurich: Welfare Agency of the City of Zurich, August 13, 1979).

Finally, the Western welfare process assumes that clients are without funds primarily because of the failure of society to provide, while the Swiss welfare agency assumes that poverty may be caused by internal or external causes and that poverty problems will not be ended until the agency and client learn the cause of poverty and do something about it.

The city of Zurich issues an annual report covering all its welfare services, including services to aged, surviving, and handicapped persons; youth services; supervision of youthful law violators and incorrigible youth; youth institutions; homes for the aged; information and referral services; and the direct welfare program (Fürsorgeamt).

The 1978 and 1979 reports (Sozialamt der Stadt Zürich 1979, 1980, pp. 288-89) list the distribution of clients served according to residency or domicile as follows:

	Number of Persons	
	1978	1979
From Zurich canton	1,110	1,140
From other cantons		
With cost sharing at 50 percent	1,297	731
With cost sharing at 100 percent	—	522
Without cost sharing	803	907
Foreign residents	347	435
Total	3,557	3,735

The 1979 report indicates that about one-third of the 3,735 clients required financial aid for the first time. Institutional welfare (in old-age homes, children's homes, and the like) represented an expenditure of Fr8,097,043. Direct aid amounted to Fr10,604,347. Both totaled Fr18,701,390. Reimbursements by relatives and others amounted to Fr11,282,812, and reimbursements by other cantons totaled Fr3,435,743, making a total of Fr14,718,555. Thus, the net cost of Zurich welfare was less than four million Swiss francs.

In the 1978 report the agency indicated that in compliance with the law relating to advancement of child support, funds for 398 children were advanced, and during 1978 another 287 cases of child support funds were registered. During this time the reimbursement coverage rose from 33.2 percent to 43.9 percent of funds advanced to clients. In addition to welfare grant cases, in over a thousand cases in 1978 and 1979, clients were given counseling and supervision with funds deriving from wage and pension payments.

The agency supervised the care of 220 children in 1979, most of whom lived out of their home. The department also operates 30 foster homes and other child-care facilities; has leased 32 homes for

mothers and children with an arrangement for the mother's lease to
expire when she marries or when the children are no longer minors;
and operates a program for 225 male and 130 female juveniles under
which children with problems owing to inadequate parental supervision
are guided in completing their schooling, in choosing an apprenticeship, and in staying with the apprenticeship until completion and job
placement. The purpose of these programs is to ensure that children
of welfare families (and other children in families who need it) will
move ahead into self-sufficiency and self-reliance. The agency describes the goal of all of its services to be a combination of inner
stability and external adjustment to employment, family life, and social responsibility. Where the parents are unable or unwilling to provide appropriate socialization and training for the children, the agency
can take steps toward court-appointed supervision, or even guardianship.

Thus, it can be concluded that the Zurich welfare agency provides much more than financial assistance. It seems to begin where
the welfare states leave off by working with the clients toward realistic self-sufficiency and social responsibility. Unlike the programs
in other Western countries, the concern is not just the alleviation of
poverty by money, but also helping clients to help themselves. Unlike the public image of welfare agencies in other countries, the
Zurich and other Swiss agencies are apparently respected by their
citizenry as competent, rational, compassionate, and concerned with
the goals of effectively helping their clients.

Below is a comparison of important data about Zurich with populations in the major Swiss cities of Bern, Geneva, and Basel.

	Zurich	Basel	Geneva	Bern
Population	367,900	181,800	161,000	145,700
Employment per 1,000	308.3	137.3	126.1	126.1
Commuters per 1,000	130.6	54.3	56.2	59.6

As shown, Zurich differs from the other three cities in terms of employment and commuters.

BERN

An examination of the civil ordinances of Bern canton in 1982
indicated that Bernese social and welfare policy is almost identical
to that of Zurich. The Bern ordinances place additional emphasis on
the confidentiality of the worker-client relationship, on the right of
clients to settle in the communities of their choice, and on the right
of clients to make their own decisions in their lives as long as they

agree to deal with the consequences of those choices. The responsibility of the client to reimburse welfare payments is outlined in instances where the welfare was secured under false pretenses, where the client is no longer a minor and is now able to reimburse the costs from assets or earnings, or where one's spouse, minor child, or aged parent has been aided. Clients are also required to reimburse if a commitment has been made to do so at the time of application. Heirs of property inherited from someone who has been aided are required to reimburse for such aid. There are also provisions for payment of interest if fraud is involved, in addition to civil penalties. If the client refuses to pay reimbursement, the welfare agency may also claim legal and court costs in securing the reimbursement.

The mandate for the agency includes the following categories of persons to be aided:

1. Needy children and young adults who require care, rearing, and/or suitable occupational training appropriate to their abilities;
2. The homeless and those living in unsuitable dwellings;
3. The needy who are unemployed, yet unemployable;
4. The needy who are sick, injured, or pregnant;
5. The destitute who require necessary dental care to prevent irreparable damage or pain;
6. The physically or mentally handicapped, if applicable, who need training for an occupation that can enable them to enter (or re-enter) the social and economic mainstream;
7. Alcoholics and those in danger of becoming alcoholics;
8. Unwed mothers;
9. Needy persons with untreatable personality defects and who are permanently unable to work;
10. Destitute people who are permanently unable to work.

The purposes of the welfare program are spelled out in the cantonal and city ordinances. The welfare agency is concerned with the well-being of people in need. The agency personnel are to listen to clients' concerns; to advise and guide them; and to help secure appropriate aid from relatives, private and public institutions, and welfare. The agency seeks to strengthen the client's self-respect, sense of duty and responsibility, and self-reliance, as well as guide the client to lead a wholesome, socially acceptable life, where resources are divided and spent carefully and responsibly. Even though only a few in the family require assistance, the agency is required to become involved with all family members. Special attention is to be given to the creation of healthy living conditions; to the operation of a clean, frugal household; and to a wholesome coexistence of family members. The agency must counsel parents regarding the care and upbringing of the children, including their educational and vocational training. If, in the agency's opinion, that care is inadequate, then the agency must provide for this to be done by a suitable person. Children and adolescents may not be sheltered in care centers or foster homes without consent of their parents, except where guardianship has been installed under the civil code.

Wherever possible, the public welfare agency must utilize the appropriate community, individuals, charities, and institutions to help secure the goals of the client. The agency must report indigents who persistently disregard instructions necessary for the client's, or client's family's, well-being. Persons in need and their legal representatives have the responsibility to add their support and energies to the goal of eliminating their need and their reliance on aid, according to the instructions of the agency. (The client's cooperation in developing a plan is encouraged, but when the client refuses or is unable to help develop a plan, then the agency may do so.) Aid is given only as long as the objective of welfare is not yet achieved (or cannot be achieved). Indispensable assistance may not be denied a needy person, even if the state of being needy was brought about by irresponsible behavior. However, if this is the case, the client is required to acknowledge the responsibility to reimburse such aid as soon as possible. If there is danger that the recipient would not use the cash assistance properly, then assistance may be disbursed in the form of coupons, consumer items, or authorizations to merchants. The agency may provide promissory notes if the needy person must supply them for such items as housing rentals, medical care, custodial care, and so forth. Neither indigents nor the persons to whom they owe money have an actionable claim to agency financial assistance.

The ordinances also empower the public welfare agencies and its institutions to demand payment from responsible relatives or clients for aid given. Voting rights may be taken away from persons who do not repay or pay reimbursible costs for themselves or their relatives. The cantonal concordat for reimbursement is authorized by the ordinances. If a responsible canton refuses to reimburse for care of persons from their canton, then the agency is authorized to deport the client to the home community. (This has apparently never occurred in the history of the intercantonal concordat.)

The client group report for Bern (Social Services of the city of Bern 1980) describes the types of services given to various client families and individuals. There are three major modes of assistance. The first is sporadic assistance given to families and individuals where the income is limited, but nevertheless suffices for covering daily needs. In such instances, payments are made for dental bills, apparel needs, a temporary supplement for living costs, and the like. The second is regular assistance. Here, a financial payment is made each month for the provision of irregularly needed supplements to income, where the income is irregularly insufficient. The third is full assistance, where no income at all is available for family support. This occurs only rarely, during a period when a disability insurance decision is being awaited or where the client is currently unable to find work, even though the client is considered employable by the insurance institution. In the Bern agency report, every fifth client was someone working full time for an income. This is because many unskilled workers are unable to provide fully for their families. The agency report lists the frequent combination of client financial problems with psychosocial and/or age or physical problems. The following are the most frequent types of problem cases reported.

The first category includes single or divorced women with children. The Bern agency reports that in 273 cases, the need for financial assistance was a result of alimony or child-support payments that were too low or irregular. In half of these cases, there were also other social problems, such as clients being relatively isolated with their children; ambivalent and problematic relationships with former partners; and women having difficulty finding jobs, making new contacts, and so forth. Often, the income from employment is insufficient for the family, and the woman has had inadequate professional training for a better position. Educational difficulties are frequent among the children, mainly because the mother is overburdened. In every tenth case, alcohol or drug abuse is encountered. There are 77 single parents who have children with serious illnesses or disturbances. In Bern, too, the agency looks forward to resolving the problem of unpaid child-care payments by use of the new law authorizing advances and repayments. The agency seeks to advise and support these families in terms of social relationships, as well as in terms of financial aid.

The second category of clients is designated as disintegrated and neglected adults. These are persons who, in prior times, would have been institutionalized. The Bern agency estimate of these persons totals 200. The basic problem of these clients is that they cannot satisfy the usual existing norms of the society. The types of specific difficulties include social adjustment problems related to alcohol and drug abuse (130 persons), criminal record and behavior (61 persons), and job adjustment difficulties (139 persons) (note that some clients are multiple listed). For about thirty persons who were unadapted and frequently discomforting to others, the Bern staff had difficulty in finding even emergency shelters. With these clients, the agency sought to preserve as much of the client's independence and self-reliance as possible.

The third client category is what the Bern agency describes as debt-ridden clients. About every second assistance client is in debt at the time of application. The bills include items taken on installment, past-due rent, and private loans. The agency does not consider third-party debts that were consummated prior to application as payable by the agency. An exception is made in regard to claims that must be paid if the client is to continue to receive vital services, such as medical insurance, electric power supply, tax debts, and so forth. Of the total, 116 clients (7.7 percent) are heavily debt-ridden. When the legal power of the creditors to effect wage attachments develops, then the agency is required to intercede with financial aid. The agency frequently negotiates with creditors and clients to seek a partial payment in settlement. In the case of 24 persons, the agency has taken over the management of their wages and pensions.

The Bern agency report of 1980 listed the problem of inadequate housing available to the aged. This is particularly grave in that there were 30,000 old-age pensioners living in Bern. There is a special need for ambulatory residential care for the aged. About 300 senior citizens were in need of nursing home care, and in 1980 the building of two new sanatoriums with 170 beds was approved by the canton and confederation. The Bern agency has moved ahead in providing insurance for all foster children under care. The 1980 report lists 2,960 cases served, in which there were 3,737 persons: 45.04 percent were assisted because of insufficient funds, 25.18 percent because of illness or handicaps, and 21.88 percent because of social disintegration.

The public welfare agencies in Zurich and Switzerland generally provide a picture of centrally and regionally located social services where the client apparently "doesn't get lost." Unlike the situation in the other Western nations, the client population is followed up, and service is given not only in terms of financial aid, but also in terms of determining client goals and then pressing through to their fulfillment. The Bern public assistance agency has a staff of 65 persons, including social workers, two supervisors, and 11 secretaries, in addition to a number of part-time social workers, institutional workers, and administrative personnel. The agency is also able to utilize numbers of community persons and volunteers in their follow-up work with the client load.

In interviews with the Zurich and Bern agencies, it was learned that the agencies hold to a firm policy of full-time employment goals for all able-bodied clients, and a part-time employment requirement for all mothers with children. Of the 3,200 estimated cases in 1983, about 1,000 were chronically dependent. Of these chronic cases, 300 are old-age retirees with certain problems that make them dependent. About 700 of 1,000 "chronics" are unemployable or less employable. The agency uses its discretionary power over the amount of assistance to help motivate client change, particularly toward reemployment. The staff adheres strongly to the principle of lesser eligibility to prevent disincentives for employment. In 1983 the agency began an experiment with young people who were former drug addicts. Where the clients might receive regular aid of Fr400 per month, besides rent, utilities, and so forth, they were authorized to receive up to Fr500 a month (plus rent and utilities) while in an employment training program. Following the training program, they can then enter into an interim job with about Fr600 (plus rent and utilities). If they succeed in training and employment, then they may move up to full market employment at Fr2,000 per month (compared with the basic aid limit of Fr400 plus rent and utilities if they fail).

The Bern agency people indicated that there is continued resistance in the schools that train social workers to the idea of grants being used as a tool for rehabilitation rather than as a floor for economic support of all. Similarly, when conferences occur among social workers of Switzerland, West Germany, and Austria, the Swiss workers are chided by their German and Austrian colleagues for not providing a standard economic aid floor for all clients. The Bern agency director also indicated that some of his less experienced workers tend to shy away from some of the realistic issues, such as the client's problems, the necessary case goals, and measures necessary to deal with client resistance to make changes toward self-sufficiency and responsibility. This pattern of worker reliance on financial assistance norms, rather than pressing for client change, is more reflective of bureaucratic civil service (frequently encountered in the other Western nations), rather than the stance of a pragmatic professional social worker.

The Bern agency provides for client appeals to the agency supervisors, then to the agency director, then to the legal offices of the city, and finally to the elected officials of the city; but in no instance is the client entitled to seek recourse in the courts, except in the case of responsible relatives who balk at paying, or in the case of reimbursements.

On the matters of "coupling" (persons living together without a legal marriage), the Bern agency takes the position that if the clients live together, then their budget and case plans are handled as if they were married.

GENEVA

The Geneva public welfare agency is listed as Hospice Général, a title that it has held since it was chartered in 1535 (one year before Calvin came to Geneva). The agency was established to bring order to the affairs of six hospitals and hospices for people in need. These were taken over as official government programs. Because social and welfare services were tied to medicine in the public mind at the time, the charter of Hospice Général is one of broad responsibility. The agency not only provides advice and financial aid but also plans fully for the care of persons with problems. Funding is by the canton of Geneva. The agency is listed as a "foundation of public right," with responsibility for all persons in need or with problems, including local Swiss citizens, Swiss from other cantons, foreigners, and "refugees" (whether valid refugees or otherwise). There were in 1983 about six thousand persons in the canton under financial support (in a population of about three hundred and fifty thousand people in the canton). Another six thousand persons were being served with counseling and advice (and did not require financial aid). The agency seeks reimbursement on the same basis as is done in other Swiss aid agencies.

There is a strict means test before any aid is granted. Unlike the law in other cantons, in Geneva the income tax report is secret and confidential and may not be examined for other than tax purposes. Because of this, applicants must make a signed declaration of their assets, liabilities, income, and so forth. Relatives are held responsible, but less so in Geneva than elsewhere, except when there are glaring instances where the relatives are known to be rich. If a client is aided and then inherits money, the agency lays claim to reimbursement and usually collects.

Regarding care of the aged and handicapped, the Geneva agency staff believe that the social insurance program is quite effective. The supplementary income program usually is sufficient for the clients, and few have requirements beyond their beneifts. Where needed, this agency provides additional aid, but primarily, the Hospice Général serves the aged and handicapped by assisting the client in making claims for the supplementary income benefits.

Regarding aid to single parents, particularly divorced women with children, the Hospice Général is opposed to substituting welfare aid for missing parental child support. This position is the same as the policy found in the other Swiss cities. The agency experiences difficulties in collection of child support primarily when the nonpaying parent is a foreigner. In other cases, husbands are easily traced via their military reserve service (or their civil defense service) or their police registration (everyone must register and deregister when moving to a new location). Husbands are easily reached by legal and financial controls as long as they have any assets or earnings. Because the cantonal government will not aid deserted or divorced women whose husbands have absconded, the Hospice Général uses its its own foundation funds for this purpose. The Hospice Général is concerned with the increasing number of Swiss women who have married black or Asian diplomatic or other visitors and who have then been deserted along with their children soon after their husbands have secured Swiss passports.

The Hospice Général report for 1982 (1983a, p. 29) indicates that "women, alone" in need represent 6.1 percent of the caseload. This was 4.8 percent of the caseload in the previous year. The agency seeks to aid such women with employment training and placement. It also limits its aid to able-bodied persons (other than the aged or invalids) by drawing up a contract with each client whereby the agency gives aid to the client for a set number of months in exchange for a promise that the client will work toward self-sufficiency. The agency declines to accept a continuing assistance plan, except in the case of aged and infirm clients. Even in the instance of young adult infirm or disabled clients, the agency prefers to arrive at an agreement in which the client moves toward employment or at least toward increased self-care. The agency is quite firm in instances where the client has not moved toward getting training or employment. Its emphasis is on getting the message across that the client must "take charge of his life and that he is expected to go to work as soon as possible" (Perrot 1983).

The agency limits its grants to subsistence needs. The principle of lesser eligibility is difficult to apply in Geneva because of the high cost of rent owing to the limited supply of housing and the heavy rate of tourist and diplomatic rentals. If a family has a low earnings base because of marginal employment and a high rental cost, the agency might help with supplementary aid if other(s) in the family would also take on part-time work. In 1983 the agency had a grant limit of Fr1,200 per month for a single person and Fr1,600 per month if there was also a child. In the case of a husband, wife, and three children, the agency could supplement up to Fr3,000 per month. But the agency could not continue such grants indefinitely. Such grants would be made only as long as the family was doing something toward eventual self-sufficiency.

The policy of the Hospice Général is to make all public aid on a loan basis. A loan is expected to be repaid in five years, but in reality, if it is not repaid by then, the agency will either forgive the loan or extend it another two and one-half years. The agency does not keep loans alive on an unrealistic basis, but it does not wish clients to believe that they are getting something for nothing. Welfare in Geneva is not a right but is given on the principle of solidarity, meaning that the community must declare its support of those in need, and those who are in need now are to provide resources for aid when they are again able to support themselves. In regard to supervision of families in their use of assistance funds and in their rearing of children, the agency is quite secure. The town and the rest of the canton is too small for anyone not to be reported, usually informally,

if they require supervision and guidance. The problem does exist
among what are called the black workers (illegally employed people
without work permits), where children remain at home unsupervised.
This problem exists only to a limited extent and usually among the
Third World people in Geneva, particularly the Turks and Iranians.

The amount of aid given to clients differs from person to person and is seldom the same month after month. Aid is adjusted to the
client's changing situation and to the client's degree of effort and movement toward self-sufficiency. No two clients are given the same
amount intentionally.

The client (or relatives who are required to make payments)
may appeal the worker's decision to the director of social assistance
and, if unsatisfied, to the executive of the Hospice Général. If still
unsatisfied, the client may then appeal to the board of the agency.
Clients who request it are provided with a statement of their rights
and the procedure for appeal. Only in the case of a responsible relative who balks at the amount required (such as an adult child in regard
to the costs of caring for parents) can the person go beyond the agency
in an appeal to the courts. There are three to four appeals per year,
but seldom is the appeal upheld. Often, the decision by the board or
court is even more forceful in terms of client self-sufficiency and
family responsibility than is the worker's decision.

Because Geneva is the location of many international agencies,
conferences, and meetings, it has become a center for refugees who
claim asylum. Thus, the agency provides the basic assistance for
the refugee families, reimbursed of course from confederation funds.
The cost per refugee amounts to about six thousand Swiss francs per
year. This is a low cost because the refugees are also given temporary employment permits while awaiting adjudication. Many asylum
applicants are refused legitimation after an investigation of two to
four years. They are supposed to be returned to their country of
origin, but by then their passports have expired. This was the case
in Geneva in 1983 for about six hundred Turkish workers. The confederation then sought to transfer financial responsibility to the canton,
but this is being argued on a case-by-case basis. More than oppression, the impetus for appeals for asylum in Geneva is the problem of
unemployment in Western Europe, which causes previously employed
blacks and Asians to seek to move to Switzerland from other European
locations. There is considerable discussion about the problem in
Switzerland because no one really wants to return people to their countries of origin if there is any danger to their lives. At the same time,
no Swiss citizens want to see their country become the haven for the
unemployed of Europe.

Because Geneva has a relatively more relaxed and informal acceptance of deviance (owing to its French-related culture and international atmosphere), many people come to Geneva from other cantons where social control is more stringent. Thus, the Geneva Canton has many more out-of-canton families who come to the local agency for help. Their costs are, of course, reimbursed by their home cantons, but they do present a heavier problem of administration and service than is the case in other cantons. The major difficulty is housing. Hospice Général has had to use some hotels because family rentals were not available. In many cases, young couples have had to stay on with their parents and double up.

With a population of 356,000 persons, 30,000 are Genevan residents. Another 30,000 are Swiss from other cantons. Foreigners make up 118,190 of the population. Of this foreign population, 43,515 are people on annual contract or annual permits, and 23,510 are daily border crossers. Out of a working population of 225,000 workers, about 2,000 are unemployed (less than 1 percent).

Regarding unmarried couples, the agency handles them as if they were married. The contract between agency and clients requires the signature of all the adults in a family.

The out-of-canton Swiss residents who come to Geneva come from areas of low employment and declining industries. Also, many German-speaking young adult Swiss come to Geneva because they seek a more relaxed and exciting atmosphere than they have in their home communes. Also, unlike other Swiss towns, Geneva's tertiary employment markets provide jobs in restaurants and hotels often without requiring work permits. This attracts many of the Western European unemployed.

Hospice Général has two social workers who do nothing else but teach undersocialized young people how to clean homes, paint, garden, perform household maintenance, and so forth and then help them to find domestic employment. Applicants for work at the Work Placement Bureau of Geneva who lack a groomed appearance, training, job readiness, and so forth are referred to the Hospice Général. The agency has five social workers who work specifically on marginal worker placement.

The canton is concerned with the high costs of handling assistance clients, but it accepts the need for this as long as aid is only temporary and as long as employment is a requirement for all able-bodied clients. A problem of acceptance of refugees by the community persists—particularly with refugees from Zaire, who are sometimes viewed as people who like to sit in coffee shops and avoid work. The problem of dealing with refugees who are not Protestant-ethic oriented persists.

The agency is large. There are 440 people in all of the branches of the Hospice Général, in 350 locations. Many of the employees (70 percent) are married women who work part-time, whose children are grown, and who have had life experience in resolving family problems.

There is a close relationship between Hospice Général and the other agencies, especially the Office de Jeunesse (the counterpart of the youth service agencies in the German-speaking areas of Switzerland). This agency handles guardianship and supervision of youth and parenting. The agency has large resources, having acquired over one hundred and twenty buildings and twenty-one hundred apartments, many by legacies over its long history. The annual income is over four million Swiss francs—from its buildings and other assets.

In an analysis of this agency's caseload, it was estimated that there are almost no intergenerational welfare-dependent families. The agency believes that this is due to the stringent policies of placing responsibility on their clients and holding them to it.

BASEL

The public welfare agency in Basel is the Allgemeine Sozialhilfe. This agency assists over 1,000 families per year. In 1982 the caseload for non-Basel residents was 1,365 families, and in 1981 the total was 1,063 families. There were also another 600 or so families of Basel residents, making a total of about 2,000 families. Of the Basel caseload in 1982, 633 families were non-Swiss residents, compared with 246 such families in 1981. There were 446 cases concerning Swiss residents from outside the city of Basel in 1981, and 520 such cases in 1982. Of the famlies assisted by the agency, less than 5 percent are long-time dependents. Unmarried couples or unmarried or divorced women with children comprise 10-15 percent of their caseload. The 600 to 700 resident dependent families are multi-problem families, but only about 150 are chronically dependent upon public support.

The agency is faced with the same problems as those of Zurich, Bern, and Geneva, in that people who are unhappy or who have problems in other settings come to the larger town. The agency upholds the right of choice of settlement and handles the problem of out-of-city dependents by billing for reimbursement on the basis of the intercantonal compact. The agency has heard of a few instances where some clients have been "encouraged" to move to a big city by small community welfare offices, but these are probably scattered instances and have apparently never been verified. Most of the nonresident agency clients are only temporarily aided. The reasons for being in need, in order of frequency, are unemployment, unmanageable debts, and illness. Underlying these are more deep-seated causes, such as inadequate training and preparation for employment, inadequate ability to manage a household and budget, lack of birth control, and so forth.

Agency policy requires everyone to work except children, who are expected to go to school. After the first week of assistance, the adult clients must present proof of registration at the public employment office. Otherwise, they will be refused aid by the third or fourth visit. If clients are not sick or aged, they are expected to explain why they have not found jobs and to discuss what they plan to do. A mother with children is allowed to stay home if her husband has left the home, but she has to sign over to the agency the right to seek child support reimbursement for the funds advanced. If a mother wants to stay home and not work, her needed services to her children are verified, as is her claim of the nonavailability of a child-care facility. The onus of finding child care and employment is on the client. Direct relatives are required to pay for care, or else the agency will advance the funds and sue for reimbursement.

The agency's aid is minimal, and in no case does it surpass what the person might earn after taxes. The agency does not aid families that own homes, automobiles, or other sizable potential assets, except in cases where the clients are only marginally aided. The budgets for families limit the level of rentals paid to Fr1,000 per month. If the rent is more and the family is small, the clients are encouraged to share the apartment. The agency supervises the rearing of client children—and does not hesitate to call in the youth service agency or the police if children are not adequately reared or controlled. When a child approaches 16 years of age, the agency worker begins inquiries with the family regarding the plans for the child's career or further education. If no career or educational plans have been made, then the parents are pressed to take the child to the career planning offices and to report back to the agency on their progress.

Regarding appeals, dissatisfied clients can apply to the supervisor, the agency director, the welfare commission, and the city presidium, in that order. Recently, the agency had an appeal from an Italian nonresident who had been in Switzerland eight years on annual contracts. He had been injured on the job owing to his illiteracy (he turned on a machine without reading the posted signs; his illiteracy was in all languages except in spoken Italian). He is now 50 percent invalid, which provides him with over Fr1,000 per month. He claimed that this was not enough to live on in Basel and sought supplementation from the agency so that he could get permission to stay on in Switzerland. The decisions on all his appeals were negative because the living costs in rural areas, and especially in the Italian-speaking areas of Switzerland, are low enough for him to live on Fr1,000 per month. The decision was that no one who is not a resident has the right to stay on in Switzerland at a community's expense. The appeals board indicated that they might have decided otherwise if the client had come up with evidence of having become literate in one of the three mainstream languages and of having made some friends in Switzerland during his eight years of living in Switzerland. The client returned to rural Italy, where his Fr1,000 was adequate for maintenance.

The most serious cases handled by the agencies are the drug addiction cases. Treatment costs run from Fr120 to Fr150 per day, aside from custodial care. Fortunately, the drug-related cases are few. The problem with these cases is that the therapy is of little proven value, and the drug clients remain unable to master even the simplest of problems. The agency believes that the bulk of the clients are beset with multiple problems (debts, unemployability owing to inadequate training and job conditioning, and so forth) because they somehow slipped through the cracks of the Swiss educational and career-training programs. That the number of such clients is small speaks well for the educational and youth service programs, but those who have missed out on training create quite an expense for the local communities.

AARAU

Aarau is a smaller manufacturing town in a metropolitan area of 60,000 people. The Sozialamt der Stadt Aarau serves a community of 18,000. The caseload varies from 80 to 100 families per year. Only one family is probably intergenerationally dependent on welfare. In this family, the life-style and alcoholism patterns tend to perpetuate the family's dependency. According to the director, the reason why there are not more is because the agency policy does not encourage families to depend on welfare for more than two to three years. If they remain dependent longer, then questions are raised in the community and in the agency's commission, and the agency has to justify why rehabilitation and self-sufficiency have been delayed. Most of the canton pays its taxes without complaint about the costs of welfare, but there is one community where primarily rich people have settled, and they tend to reject all social costs because they claim that there are no social problems in their community. These people do pay their share of the taxes to the canton, but there are some complaints.

When people move within a canton (such as Aargau) and become dependent, the intercantonal compact does not apply. Instead, an intracantonal compact is usually developed by the towns in the canton. In the case of the Aargau intracantonal contract, costs for a dependent family are paid by the original commune for the first two years of settlement and then by the community of settlement.

The agency policies are similar to those of other agencies. The means test is fully enforced. Some of the social workers are reluctant to check their client's assets, but in Aarau, they are required to do so, nevertheless. The agency requires a check of the client's income tax report, wealth tax report, and other relevant documents. The financial aid is a loan, not a grant. This is particularly so with younger people, in order to prevent the development of a pattern of dependency. Aid is always granted on a contract basis that includes arrangements for reimbursement. Reimbursement is the established policy in all of Switzerland, but in Aarau this is strictly enforced, except when hardship would result from the reimbursement process.

If a young mother wants to stay home with her young children and not work, then the agency may give her a regular grant, but only until the children are old enough to leave in a day-care center; then she is expected to go to work. If further pregnancies should occur that might delay her return to work, the agency would press for proof of parentage and then press the father to provide support. As it is, women with children must cooperate with the agency in securing child support from the fathers—otherwise, they will be refused aid. If the mother finally goes to work and earns just enough to support the family, then no attempt would be made to seek reimbursement. But if the mother comes into money or gets a very good job, then repayment would be considered and discussed.

Regarding employment, the agency expects the client to work under all circumstances. If the client has no work, the agency will work with the employment bureau to help find a job, even if it is unskilled or marginal. The client cannot expect to be picky about work while using the agency for support.

Regarding grant amounts, the agency reserves the right to give different grants to clients based on their different needs, readiness to secure and keep employment, and readiness to work with the agency on a cooperative rehabilitation plan. The trend of individuation is more evident in Aarau than it is in Bern or Zurich, according to the agency director. The agency has to help the client change. Emphasis is not to be kind, but to be effective. Workers are allowed to use discretion and may give less or more aid, using the grant as motivation for client change.

The agency provides for appeals, first to the director, then to the city council. Where the decision relates to the payment for a member of the family, or where it is a repayment matter, the appeal may even go beyond the city council to the courts.

The agency abides by the laws and ordinances passed by the city council. In general, the agency aids families based on the amount required to subsist, rather than the amount required to match the social existence level of the community. Thus, the family can count on rent, food, utilities, and insurances but not on a car or travel holidays. If a child is in a drug rehabilitation program or a psychiatric facility and if the parent shows total subsistence needs of Fr2,000 per month and earns Fr3,000 per month, then the difference of Fr1,000 is divided by 3, and the parent is expected to contribute between Fr300 and Fr350 per month for the care of the child in the drug cure or psychiatric facility. The balance would be supplied by the agency.

In divorce cases the judges usually levy a child support payment on husbands with a minimum of Fr300 to Fr400 per month and a maximum of Fr600 to Fr750 per month per child. Usually, judges do not grant alimony. In the few instances when it is granted, it is usually for Fr400 to Fr500 per year, and then only for one or two years during the postdivorce adjustment period. When the husband fails to pay, the agency will pay in his place and then lay a garnishment on his wages. The agency estimates that about 30 percent of the agency caseload represents divorced or unmarried mothers, about the same that is reported in Bern and Zurich. Another 30 percent of the caseload is made up of single adults who drink or take drugs. These clients are less employable. Mainly, these are males, but some women are now appearing in this problem group. The agency staff believes that it is the agency policies and local control that tend to keep welfare delinquency within limits in Switzerland. They also believe that the informal community social controls support and reinforce the limits on welfare dependability. One of the negatives of such social control is a lack of warmth between people. Individual responsibility reinforces the barriers between people. Thus, there is a greater social distance between parents and children or between friends or acquaintances. This leads to generation gaps and loneliness.

Another negative aspect is that emphasis on individual responsibility leads to a higher suicide rate and a higher rate of ulcers, colitis, and neurosis.

Regarding local control, each case is written up and presented for review to each of the seven members of the city council. Each member indicates which cases should be discussed by the council. If all seven vote "no discussion," then the case is processed by the staff without change. If one asks for discussion, the case is then discussed and must be reviewed further whenever a new or increased grant or a new and sizable expenditure is planned. The agency's experience has been that the city council members are quite paternalistic, but they keep their distance and remain quite uninvolved in the client situations, especially when expenditures are involved.

The agency's four areas of responsibility are the local welfare service, including case aid, children's home contracts, and so forth; juvenile delinquency control and legal measures for guardianship of the young, sick, aged, and the like; the community youth services, including the community youth center; and kindergartens, child-care centers, and children's overnight centers under contract.

There is also a youth advice agency, which is funded by a cooperative arrangement between the Protestant Evangelical Reform church and the Catholic church. This agency informally works with the welfare agency whenever a legal matter arises, such as guardianship of a child or a follow-up on an absent father.

In addition to the 80 families on the regular caseload, there are about 20 to 25 refugee families supervised by the agency. Any expenses incurred through aid to refugees are reimbursed by the confederation. Currently, all the refugee families are employed and self-sufficient.

The three most difficult cases in the agency were discussed. One involves a locksmith who is very capable but an alcoholic. He is 26 years old, as is his wife. They had an early marriage and they have children, ages 2, 3, and 4. The family came to Aarau from another community in Aargau Canton. The agency's diagnosis is "work shy," in that the father avoids work. The family has been on welfare since 1978 and has been a heavy cost to the community since 1980. The wife is psychiatrically disturbed, with some brain injury and impairment but no external disability apparent. She is somewhat overweight, defensive-aggressive, and angry at the agency for seeking to get her to better order her family's life. The husband has been an alcoholic for six to eight years and is generally labile in his patterns. The situation includes problems with birth control—which the agency is not allowed to press; moderate family disorganization, but without child or spousal abuse or neglect; growing family dependency on the community; and inadequate role models for the children. The agency is planning to provide day-care facilities for the children in the future, in the hope that they will be helped to learn self-sufficiency from external role models. The agency also plans to secure psychiatric consultation if no resolution is forthcoming.

Another case concerns two intelligent parents and involves alcoholism and divorce. In this instance the woman is alcoholic and the man is "diffuse" in that he flips from subject to subject, likes to "play doctor," likes to play with ideas, and takes nothing seriously. Both are native Swiss. Five years ago the father went to Lebanon, but what he did there nobody knows. He is now age 45, and she is 47. They have one child, age 13. The father gets money by deviant means, such as payments from a former girlfriend in Lebanon. The mother has been in and out of psychiatric hospitals and alcoholic treatment centers. The child was taken from her after many alcoholic bouts and is now in a children's home with a structured program as prescribed by the child psychiatrists. The mother wants permission to visit more frequently than once a month for a day at a time, but the therapists report that she disturbs the child. The mother's own therapists want the mother to have more time with the child. A possible solution might involve a case conference of the two sets of psychiatrists. The agency is also considering working out a contract with the mother, setting up steps of achievement for her that would be rewarded with more time with the child. These achievements would be related to alcoholic controls, employment preparation, and so forth.

The third case involves a divorced couple and five children. The oldest child, a young woman, is now 20 years old. She was incestuously attacked by her father six years ago. The divorce occurred after that. The father is now remarried. He makes child-support payments regularly, but his payments do not cover the children's full costs and he really cannot afford more. Because of the disturbed condition in the home, two of the children are in institutional care (including the incest victim), two are in a foster home, and the youngest—a boy—is in a children's psychiatric unit. The boy, age 13, wants to be in contact with his father. All the children want to see each other and are brought together for occasions periodically. The mother is extremely aggressive and overactive. At times she disappears. The agency has guardianship over the children but would like to bring the mother and children together on a more stable and cohesive basis and is working toward this goal.

All three cases represent a breakdown of family structure under the impact of individual failure. They also represent a failure of the extended family to help when the nuclear family falters. Each instance demonstrates the ease with which families can push off responsibility onto the community and how this process moves from smaller to larger communities unless the community and its agencies seek to prevent it.

Regarding absent fathers, the agency believes that this is not too great a problem. If they are Swiss, they can always be found, either through the community registry lists, the tax rolls, or the military reserve lists. Only if they leave the country do difficulties occur. The real problem with absent fathers occurs mainly in the case of non-Swiss who leave the country. Swiss fathers, even when they leave the country, must pay 3 percent of their income to support the military budget (in lieu of not serving in the reserves during their absence abroad). Even if the male is not employed and does not have any income, he must still pay a minimum tax of Fr200 per year or else serve time in prison. Only in the case of foreign fathers who leave the country is there no recourse. A Swiss consul has no authority to press foreign workers in their own country. The only recourse in such cases is attachment of the old-age, survivors', and invalid insurance payments that the foreign father may have earned while in Switzerland, but these payments may be limited.

The agency in Aarau suffers in that it lacks the specialized facilities available in the larger towns, such as "hot lines" for youth and shelters for abused spouses or runaway or abused children. The private or voluntary agencies in Aarau are also limited in specialties and resources.

Regarding the use of volunteers in service to families, the agency has found that they require careful supervision. Lay workers and case decision makers tend to be oversympathetic at the beginning and then restrictive and punitive when the family remains overdependent. The reverse occurs with professionals, who become so understanding of client problems as to be tolerantly permissive and tend to trap the client in a "box" of dependency on the agency. The agency head believes that the use of lay council members, as well as professional workers, provides a necessary balanced reality.

In the case of divorced fathers, usually the divorce judges cannot prevent them from remarrying. But when the judge calculates the mandatory child-care payment, it is often assumed that the new wife is able to work and will do so—at least part-time. The judges are quite restrictive of fathers. Many divorced men think that this is not just, but community attitudes support the judges' position generally.

On the other hand, a mother who wishes to remain at home, rather than work at least part-time, has to have a good reason for it. Otherwise, she will not have the support of the judge and agency. In the last ten years, an adult school has developed in Aarau (open afternoons and evenings) to help women (and others) develop careers or apprenticeships and supplement earlier educational achievements.

SUMMARY

A general description of public welfare programs and expenditures has been presented as well as details of specific programs in four larger towns and one smaller town. In each the local unit of government is responsible for care of the needy, both professionally and financially. In each town the program is individualized to meet the client's particular problems. The client group is provided with aid on a loan basis, or at least on a reimbursible basis if repayment is possible. In each town the client is held responsible to help develop and follow up on a rehabilitation program. Clients are not only aided but supervised in their rehabilitation. Except in the case of the aged and infirm, no client is given more aid than can be earned—and thus the incentive to be employed is ensured. The agency, in each instance, is carefully overseen by elected civil officials who are, in turn, held responsible by the electorate of the community. All welfare is temporary, and families are held responsible for the care of their needy members.

In conclusion, the number of temporarily dependent clients is minuscule when compared with the caseloads in cities such as New York, Los Angeles, or London. Unlike the situation in the United States, the number of intergenerationally welfare-dependent in Switzerland is so low as to be difficult to calculate. The welfare system in Switzerland, unlike that of other Western nations, is effective. In its approach to the control of welfare-dependent families, it is ultimately more humane and compassionate than other Western nations, where children are born into "welfare-trap" families fostered by the very programs that supposedly exist to help them.

REFERENCES

Büchi, O., A. Inglin, A. W. Stahel, and P. Tschümperlin. 1979. "Die Organisation des Sozialwesens in der Schweiz" [The organization of social policy in Switzerland]. Report of the three-nation student assembly, Frankfurt, November 19-23.

Charles, Jean-François. 1981. "Non utilisation et abus des services et préstations en matière de sécurité sociale" [Nonutilization and abuse of services and applications in matters of social security]. Schweizerische Zeitschrift für Sozial Versicherung (Bern), 25th year, vol. 2.

Hauser, Jürg A., Franz Bluntschi, Rudolf Hohn, Elizabeth Monig, and Lorenz Wolfensberger. 1980. Empirische Aspecte der Fürsorgebedürftigkeit: Am Beispiel des Kantons Zurich [Empirical aspects of public welfare need: an example of the Zurich Canton]. Bern: Paul Haupt.

Hospice Général. 1983a. Hospice Général rapport annuel, 1982: Institution genovoise d'action sociale. Geneva: Hospice Général.

_____. 1983b. Politique d'assistance et d'aide sociale. Geneva: Hospice Général.

_____. 1983c. Rapport annuel, 1982. Geneva: Hospice Général.

International Labour Office. 1983. The Cost of Social Security: Basic Tables, 1978-80. Geneva: International Labour Office.

Keller, Theo. 1982. "Grundzuge der öffentliche Fürsorge" [The main feature of public welfare work]. Zeitschrift für öffentliche Fürsorge 79: 66-75.

Los Angeles County, Department of Public Social Services. 1984. Fact Sheets, July 1982-June 1983. Los Angeles: Los Angeles County, Department of Public Social Services, Research and Statistical Division.

Newspaper Enterprise Association. The World Almanac and Book of Facts, 1985. 1985. New York: Newspaper Enterprise.

Perrot, Guy. 1983. Director of Hospice Général. Interview with the author.

Social Services of the City of Bern. 1980. Klientengruppen und Problemsituationen: Kommentierte Statistik, 1980 [Client groups and problem situations: commentary and statistics, 1980]. Bern: Beilage zum Verwaltungs bericht der Fürsorgedirektion.

Statistisches Jahrbuch der Schweiz, 1982 [Statistical yearbook of Switzerland, 1982]. 1982. Basel: Birkhauser.

Strang, Heinz. 1970. Erscheinungsformen der Sozialhilfe bedürftigkeit: Beitrag zur Geschichte, Theorie, und empirischen Analyse der Armut [Visions shape social aid to the needy: contribution to the history, theory, and empirical analysis of poverty]. Stuttgart: Ferdinand Enke.

Street, David, George T. Martin, Jr., and Laura Kramer Gordon. 1979. The Welfare Industry: Functionaries and Recipients in Public Aid. Beverly Hills, Calif.: Sage.

Urner, Paul. 1979. Grundsatze des Züricherischen Fürsorgerechts [Principles of public welfare law in Zurich]. Zurich: Welfare Agency of the City of Zurich, August 13.

6

INDIRECT
CONTROLS OF
WELFARIZATION IN
SWITZERLAND

THE SWISS FAMILY

Unlike the condition of the family in many other Western countries, the position of the family as a functioning institution is robust
in Switzerland. The Statistisches Jahrbuch der Schweiz, 1982 (1982,
p. 577) lists the comparative divorce rates for 1979 for Switzerland
and other countries. The highest national divorce rate was 5.3 per
1,000 inhabitants for the United States, followed by 2.5 for Denmark
and 2.4 for Sweden. Switzerland's divorce rate was 1.6, comparable
to 1.7 for the Netherlands and Austria, and lower only in Belgium at
1.4. Swiss marriages in 1980 were 5.6 per 1,000 inhabitants, and
the comparable figures in that year ranged from 4.5 in Sweden, 5.2
in Denmark, and 5.4 in Norway to 5.9 in West Germany, 6.2 in
France, 7.5 in England, 6.0 in the Netherlands, and 10.4 in the
United States. Thus, it appears that marriages occur at a lesser rate
in Switzerland than in many other countries, but then, so do divorces
(Statistisches Jahrbook der Schweiz, 1982, p. 581). In recent years,
Swiss marriages have been occurring somewhat later in life, and the
divorce rate has leveled off. The Swiss problem of divorced families
with children can thus be considered less serious than in other Western countries.

Regarding children born out of wedlock, the Swiss data show
similar developments. Hoffmann-Novotny (1983) indicates that the
number of children born out of wedlock per thousand live births is 4.9
for Swiss citizens and 6.3 for foreigners in Switzerland, making an
average of 5.2 for the nation in 1981. The data on the United States
(U.S., Congress, House, Select Subcommittee on Children, Youth
and Families 1983) show a 1980 rate of 18.4 percent or 184 per thousand live births. The U.S. rate is therefore 36.8 times the Swiss

rate. The white illegitimacy rate in the United States is 10 percent, or 20 times the Swiss rate. The nonwhite illegitimacy rate in the United States is over 50 percent, or 100 times the Swiss rate. Furthermore, the Swiss illegitimacy rate is offset by postbirth legitimation of about 3,000 children per year, usually during the first year of the child's life. Out of 3,801 children born out of wedlock in 1981 (Statistisches Jahrbuch der Schweiz, 1982), 3,165 were legitimated. One can therefore conclude that the problem of single-parent families is not usually a problem of unmarried motherhood but, rather, exists primarily in families after divorce. Even in this regard, Switzerland has much less of a problem than in other countries.

Unlike the situation in the United States and in many other Western nations, the unmarried-motherhood rate is kept low in Switzerland by social pressure, parental authority, and by the functioning of most local Swiss ordinances under which unmarried parenting is considered a violation of the civil code. Thus, the pregnancy of a girl out of wedlock may lead to a declaration of the new parents as law violators subject to Vormundschaft ("the appointment of a legal guardianship"), especially if either of the parents is a minor. Under such circumstances, Swiss adolescents and young adults are motivated to avoid pregnancy. If it does occur, the next step is usually marriage, legitimating the child often before the child is born.

The 1980 census as reported by Lüscher (1983) indicates that about 12 percent of the family households with children had only one parent present. These one-parent families are less the result of widowhood and more the result of the higher divorce rates experienced in the 1970s. In 60 percent of the Swiss families, there is only one child. Lüscher concludes from the data that frequently there is a new partner in the family with whom a marriagelike arrangement exists. It is believed that about one hundred thousand people live in single-parent Swiss families. The percentage of all couples living out of wedlock is almost 8 percent. Here, too, Lüscher believes that these consensual partnerships in Switzerland, more so than in other countries, are merely a phase before the marriage ceremony and the rearing of children. Lüscher compares the percentage of young women (ages 20-24) living in a consensual partnership at 7.9 percent with similar French, German, and Norwegian rates of 12 percent and Swedish and Danish rates of 30 percent. He indicates that the rate for German-speaking Switzerland is 7.5 percent and that the rate for French-speaking Switzerland is 11.2 percent. The rate of consensual couples with children in Switzerland is substantially lower, less than 2 percent of all families in Switzerland with children. In the whole of Switzerland, only about twenty thousand households exist in which there is an unmarried couple with children. Lüscher concludes that the intact family with children predominates in Switzerland, probably more so than in other Western nations.

It may be that the lower rate of marriages is a result of the greater care and preparation required for a Swiss marriage. In Switzerland more so than in other countries, marriage concerns not only the young couple but also the two extended families. Similarly, Swiss young people are less apt to get married before the groom is at least well established economically. Delay in marriage is also a positive factor in restraining the divorce rate because the couple is more mature, and this helps ensure a more stable family life.

There are differences in the proportion of marriages to divorces in the various locations. In Fribourg Canton, for example, there are 18.7 marriages for each divorce (Switzerland, Federal Office of Social Security 1978, p. 21). In Obwalden Canton there are 11.6 marriages for each divorce. The weakest ratio (2.9) is in Geneva. Zurich has a 3.52 ratio and Basel has a 3.18 rate. Many of the other Swiss towns range between Zurich and Lucerne, which has a 7.18 rate.

The U.S. ratio can be calculated from the The World Almanac and Book of Facts, 1985 statistics (Newspaper Enterprise Association 1985), showing 2,444,000 marriages and 1,179,000 divorces in 1983. This U.S. ratio is therefore 2.07, making it less than that of Geneva (2.9).

Luck (1978) describes the picture of the Swiss family making ascents together over the Wanderweg ("hiking path") built into the sides of many Swiss mountains. The Swiss family spends much of its leisure time together either at home or in active recreation. The cohesive bonds both within the nuclear family and throughout the extended family are reportedly robust. In fact, the closeness of the Swiss family has been criticized by outsiders (American Women's Club of Switzerland 1982; Kubly 1981). Schmid (1981) speaks of the cohesion of Swiss families, both nuclear and extended. Professor Tuggener of the University of Zurich, in his description of the Swiss social condition during an interview with this author, indicates that part of the stability of the Swiss family and community is the result of an undisturbed social structure inherited from pre-World War I times, or at least from World War II. In this kind of structure, which has nevertheless provided some permeability for movement within it, there is at least a place and a role for everyone in the nuclear and extended family and community. By careful and thorough socialization provided in the home and in school, people learn what roles and positions are open to them as well as the approved methods by which they can achieve their places in the society. A father would hardly run away, or even try to, as long as he is known both locally and regionally. Similarly, a mother would not think of leaving her family or "pushing out" her husband without very important reasons in a community that she knows observes her and that she can expect will continue to observe her. In Switzerland one is infomally answerable not only to the neigh-

bors but also to the people "back home" in one's community of origin.
If a separation or divorce occurs, both involved are answerable to
friends and neighbors to a greater extent than elsewhere. This is unlike life in the U.S. community, where people mind their own business. In the United States it is considered rude to ask a divorced
person what happened to the seemingly ideal marriage. This is a result partly of the ethos built up during the 1970s, emanating from a
social revolution associated with the women's liberation movement
and the "me" decade. These developments have had little significance
in Switzerland.

Because the Protestant ethic is very much in evidence in Switzerland, the rearing of children in the family is considered critically important. Learning is viewed as the job of the child, just as supporting
the family is viewed as the primary job of the husband, and just as
keeping and managing the home is viewed as the primary job of the
wife. The child's learning tasks are seriously undertaken, first at
home in the family and then at school. Even after the child begins
school, the Swiss parents take a strong interest in the child's learning progress, more so than in many other countries. Parents frequently ask to see their children's homework before allowing them to
play. Just as failure on the job is considered more serious than it
is in other nations, so does failure to learn become a matter of concern not only for school authorities but also for members of the family, both nuclear and extended. Unlike the U.S. educational view,
which allows children to "grow into learning," the Swiss view is one
of responsible children meeting preestablished standards and prescribed assignments. In many areas of Switzerland, proficiency is
required in specific achievements before the child is allowed to move
on to the next class.

The principle of subsidiarity is taken seriously by the family
in Switzerland. Nothing is relegated to the community or to community agencies that the Swiss family believes belongs to it. Loyalty
to family, to community, to canton, and to nation are instilled in family members, and in that order. The family has a deep concern for
what others think about it. This Swiss version of the Golden Rule
shows a strong sense of responsibility for the effect of one's behavior
on the rights and life of others. Examples of this emphasis are diverse. Under most cantonal laws, a family head is responsible for
the behavior of the children. This means that reimbursement is required for all child vandalism, graffiti, or damage to the property of
others. Most parents carry a special insurance policy to protect
them from claims, but with a deductible amount paid by the parent.
The rates are very low because so little damage is done by Swiss children. Another example of this concern for the welfare of others was
the cancellation mark used over stamps by the Swiss Post Office in

1983. It said, "After dark, consider your neighbor—Keep radios and TVs at low volume." Unlike the contemporary U.S. pattern, which promotes "letting everything hang out," or "if it feels good, do it," Swiss rearing of children encourages careful consideration of and respect for others and for the property of others as well. This starts with respect for and adherence to the wishes of elders and continues on to respect for and adherence to the wishes of one's peers. Children are taught that they must carry individual responsibility for, and suffer the consequences of, their actions. Parents teach their children that their role is to learn and be helpful, rather than "to enjoy"— seemingly the U.S. imperative for its young. The Swiss children are taught early in life to be self-reliant, responsible, and committed to the norms of reciprocity.

In families in many parts of the world, parental support of the child is provided along two lines. The first, usually associated with the ever-supportive homemaker mother, is an unqualified acceptance of the child, regardless of behavior or competency. The second, usually associated with the critical father (who is also the main family provider), tests the child in behavior and competency, praises that which he finds praiseworthy, and calls attention to shortcomings in terms of assignments for improvement. This dual pattern of parenting can be found in various historical and cultural records. With the entry into the employment market of many mothers, their desire to perform successfully on the job may influence and shape their children accordingly.

It can be hypothesized that some of the problems in U.S. education derive from the sizable proportion of families with only one parent present (mainly the mother) and from the equally sizable proportion of families where fathers are frequently absent from the home because of occupational or personal conditions. This is not the case in the smaller proportion of Swiss single-parent families nor in the Swiss cultural pattern, where the child is answerable to the father and where the father's position in the family is strongly supported by the mother. The father's position is further strengthened by the extremely low unemployment rate in Switzerland, which reinforces the father as primary family provider.

The Swiss family structure, in its present condition, is strongly supportive of the preparation of academically competent, job-ready, self-sufficient youths who enter the job market in the steps of their employed paternal (and, in many cases, maternal) role models. Thus, unlike the patterns believed to exist in many other Western countries, the Swiss father (and, increasingly, the Swiss mother) may be seen to serve as "coaches" and examples for children preparing for self-sufficiency.

In Switzerland, the family as an institution is supported not only by the culture of the society but also by the local community in which it is located and by the community institutions. Somehow, in the towns, communities, and even local districts of the one larger city, the parents are known to most of the community residents and officials.

There are, of course, some serious exceptions, as was reported by the social workers of the Youth Authority (Jugendamt) in the Horgen Bezirk District of the Zurich Canton and of Bern. In some of the former farming communities, now recently developed to provide housing for commuters, many of the leaders of the community do not yet know all the new residents. As a result, some new families are still "lost" or "lonely" and have problems that have not yet been examined by the community. We believe, however, that this problem represents only an interim experience, which the new communities are now addressing. Other problems include the high cost of land for house plots, which may require the introduction of more cooperative apartments and multiple units, as in the new city developments in the United States and the United Kingdom. A positive factor in the support of Swiss family life is the position taken by most public social workers in the community Youth Authority and the community welfare bureaus, where the workers involve themselves in family activities rather than wait for the clients to request help. In a sense, the Swiss social work stance represents a balance between advocacy for the client and support of societal standards and expectations. In many other countries, social workers have done less of the latter and, instead, are helpful to the client only when called upon.

It is true that Switzerland has some problems with the housing supply that creates difficulties for the beginning family. And yet the problem is not as severe as in other countries in many respects. In the United States, not only is land around cities more expensive but interest rates make mortgages expensive for a beginning family. In Switzerland public transport is not expensive and is, in fact, quite convenient, to the degree that families can move into the hinterland and commute without too much difficulty. In addition, because of the credit situation, a mortgage on a Swiss home can be secured at about 5 percent, and many banks do not insist on payments on the principal as long as a suitable down payment has been made. More to the point, with many extended families in robust condition, there is help for the beginning family from close relatives.

That the family is expected to continue in Switzerland is obvious when comparing the percentage of people never married by age 50 in the period 1870–1970 (Switzerland, Federal Office of Social Security 1978). Never-married men were 21.2 percent of the male population in 1870, but only 9.8 percent of the male population in 1970. Never-married women went from 22.2 percent of the female population in 1870 to 12.9 percent in 1970.

In terms of number of children, the same source (p. 38) indicates that Swiss family patterns were as follows (in percent):

	1951	1963	1972
No children	24.1	16.6	30.0
One child	16.9	9.0	15.4
Two children	25.1	27.7	34.4
Three children	15.9	22.7	14.1
Four children	9.0	12.4	4.3
Five or more children	9.0	11.6	1.7
Total	100.0	100.0	100.0

Thus, of the 1951 families, 57.9 percent had one to three children; for 1963 it was 59.4 percent; and for 1972, 63.9 percent. It appears that more Swiss families are limiting themselves to three or fewer children. This pattern is similar to that of most Western nations in which the middle class has fewer children, thereby ensuring effective socialization, education, and socioeconomic mobility.

Another factor indicating family stability is the age distributions of wives in the families. In 1960 only 0.2 percent were 19 years old or younger, and in 1970 only 0.4 percent were in that age group. The rest of the wives in Swiss families (99.8 or 99.6 percent) were 20 years or older at the time of marriage. Almost 95 percent of the Swiss wives in 1960 and 93 percent in 1970 were over age 25. Because of the very low adjusted rate of unmarried motherhood (after legitimations) in Switzerland, this would indicate that children are primarily born to mothers over 20 years of age, and most are born to mothers 25 years old or older. The comparison with the United States is made by Fuchs (1983), who lists a teenage fertility rate of 52 per 1,000, a rate five times that of Switzerland. Thus, the Swiss family pattern is for most children to be born in families with both parents and with a mother over 25 years of age. The U.S. pattern is for sizable numbers of children to be born to teenage unmarried mothers and for proportionately less children to be born in families with both parents and with mothers mature enough to be able to participate in effective socialization of the children.

In addition to a stable family structure, Switzerland's communities, cantons, and the confederation supply a variety of other supports for family life. In each of the communities, special low fares are provided for family trips on local transit systems, railroads, buses, and lake steamships. Special low-rate school passes are sold for children and youths attending schools or training. Stipends are provided by each canton in established amounts for the particular level of training. A total of 54,500 stipends were distributed in the amount of Fr140 million in 1976, making an average of Fr2,573 per student for the year. Stipends granted to students need not be repaid. They are based on each family's financial situation and on the type and cost of the education undertaken by the student. Cantons also grant loans to students that are to be repaid after completion of the educational training, according to a previously agreed-upon time schedule: Fr19.2 million were loaned in 1976. The loans are not subsidized by the confederation, but the stipends are partially subsidized by confederation funds. Unlike the U.S. educational assistance plans, however, the canton is the ultimate authority in determining how much of a grant will be given. This is based on the belief that the local community or canton knows much more about the needs of its members and of their ability to supply their own needs (Switzerland, Federal Office of Social Security 1978).

The cantonal laws concerning student stipends are based on the principle of dependency on parents for an education. The parents' primary obligation is to train their children for a livelihood. This obligation includes paying for not only the student's direct living needs but also incidental educational costs. When parents cannot carry the full costs (based on the principle of subsidiary support), then stipends and loans are provided by the canton, based on the family's economic situation and the child's potential talent. Under cantonal law, children are required to conduct themselves according to the wishes of their parents and the school authorities if they are to continue to be eligible for their stipends. The guarantee of stipends makes the cost of children's education considerably lighter for the parents.

The amount of the stipend granted is calculated by a point system in many cantons. The lower the parental income, the higher the points granted. Additional components of the point system are the number of other dependent siblings, the foreseen educational costs, the additional costs if education is located away from home, and the foreseen or actual earnings of the student.

Only in Geneva is there a claim prerequisite for all students. Here students, on the basis of their families' tax assessment, may secure a stipend without application. In Geneva it is up to the student to decide whether or not to make use of the available stipend. Geneva Canton has chosen to make the stipend a right rather than a matter of available aid based on need.

The range of stipends based on the level of institutional education is as follows (in Swiss francs):

	Average Minimum (for most cantons)	Average Maximum
University attendance	530	7,600
Technical school	530	7 500

College preparatory, teachers' college, career training, training for home economics, vocational schools, professional schools, and continued education above and beyond professional education are provided with approximately the same minimum and maximum stipends, except in instances where schooling is available in the same community.

The number of stipends provided in each canton is generally congruent with the cantonal population size.

Another major aid to families is provided by cantons in the form of tax exemptions offered to families with children. Zurich provides a deduction from tax of Fr9,000 for a couple with children, plus Fr1,800 for each child. Basel provides a Fr2,000 deduction for each child. Bern provides a married couple Fr4,800 plus Fr1,000 per child, plus a 10 percent deduction of taxable income (minimum, Fr1,500; maximum, Fr2,500). If a child is being educated away from home, the deduction is raised to a maximum of Fr1,800, dependent on the location. Similar deductions are provided by other cantons. In addition to these, each of the cantons provides deductions for insurance premiums paid out by families with children.

Besides tax deductions, each canton provides limited loan guarantees for family homes (a program similar to the Federal Housing Agency mortgage insurance in the United States) and grants for home improvements. There were 16,862 such grants provided between 1951 and 1977, amounting to something over Fr116 million. In addition, there were over 40,000 grants for housing provided by the confederation through the cantons, amounting to approximately Fr590 million from 1966 through 1976.

There are also special family allowances provided to agricultural workers and small farm holders. In 1977 there were 6,027 allowances to agricultural workers, 4,487 supplementary allowances to farm households, and almost 11,000 supplementary allowances to farm children. For small farm holdings, there were 27,000 family allowances, plus almost 80,000 farm children's allowances in 1977. These allowances totaled something over Fr138 million in 1977. Under the 1974 amendments to the Agricultural Family Allowance Law, each agricultural worker's household receives Fr100, plus Fr50 for each child (Fr60 in the mountainous areas). Eligibility for such family allowances is based on an income less than Fr16,000 for the family, plus Fr1,500 per child. In addition, there are supplements for farm family allowances in eight cantons. There are also sizable family allowances for workers in general in each of the cantons. These allowances range from Fr60 to Fr85 per child per month, dependent upon the canton. In addition, five of the cantons provide educational grants for children averaging Fr100 per year per child and grants of Fr200 to Fr500 per child at birth.

The family allowances are funded by an employer's payroll tax ranging from 1.25 percent to 2.50 percent, according to the cantonal electoral decisions. In addition, each of the cantons provides a monthly supplement to foreign workers in Switzerland at a rate of Fr 50 to Fr 85 per child for children under 16 and, in many cantons, for nonemployable children up to age 25.

It can be seen from the above data that the Swiss society is concerned with support of the family. Each of the laws and regulations in the cantons and the Swiss Confederation is periodically reexamined and amended to fit changing conditions. The laws relating to agricultural family allowances, for example, have been amended nine times between 1944 and 1974.

In addition to the above measures, each of the cantons and most of the communities also operate family and individual guidance programs, most of which are integrated in the Social Service Office (Sozialamt), which also provides welfare services. These offices are in addition to the Youth Authority, which is separate from the Juvenile Control Office (also under the Social Service Office), which provides a service similar to juvenile courts and juvenile probation in the United States. Each of the Social Service Office bureaus is located in the towns and in regional and neighborhood offices. Because the Social Service Office is directly responsible for aid to families in the resolution of problems and in the prevention of serious family problems, the staff and elected officials seek out the families and individuals "at risk" and usually make vigorous efforts to reach them before they are in difficulty. There is a continuing effort toward career direction and the prevention of unemployability, alcoholism, drug addiction, and other social ills.

In addition to these resources, many employee groups and employers provide others. Workers for the Swiss Federal Railroads are provided with a loan fund, an aid fund, a housing aid program, and social services. Workers for the Post, Telephone, and Telegraph System are eligible to participate in a broad-ranging welfare fund and a mother's vacation program. Many private employers provide similar services.

Thus, probably the most effective of the informal controls on welfarization is the Swiss family structure and its support services. From an early age, family socialization at almost all levels of social stratification operates to shape children into functional citizens and workers. Community attitudes reinforce the authority of the family over children, and children soon learn that their behavior must conform to a standard of consideration for others. Children are impressed with the assignment of picking up after themselves, of doing the work assigned to them, and of preparing themselves for self-sufficiency. Children soon learn that, as members of a family, they must care for their siblings and be responsible for their well-being. Because of the effective family socialization, few individuals find themselves in need of community aid.

THE SWISS COMMUNITY

Still another reinforcement comes from Swiss community planning. After World War II, the planning of communities, the location of factories, and the construction of residential development were based on a policy of modern industrialization without urbanization. This was not owing to national planning but was the result of strong local initiative and decentralized political power. Aside from Zurich (often called the world's largest village), which is not a megalopolis when compared with other metropolitan areas in the world, there are no large cities in Switzerland.* Most of the population live in small villages and towns, which are easily reached by an effective network of suburban trains, trams, and buses. The residents of these towns may work elsewhere, but they participate actively in communal decision making. Many communities function on a town hall basis, and most decisions regarding expenditures, schools, housing, and welfare policies must be ratified either by an election or a meeting of all electors. The social distance in these communities is minimal and news about individuals travels fast. In this way people hear about jobs, informal community help is given to the aged and sick, and there is a high level of informal mutual aid. In such a setting, a considerable degree of neighborly aid is frequent, as is the operation of numerous voluntary social service programs. Also, in such a setting, public dependency is minimal and people are more readily helped to settle their problems of employment and economic independence.†

---

*The city of Zurich has approximately 354,000 people, and the Zurich urban area has about 720,000 inhabitants. Thus, Zurich is itself a small city in comparison with the metropolitan centers of the world. Additionally, Zurich and the major Swiss towns are losing population to Swiss villages, which are growing into self-contained communities. Basel is 49 percent the size of Zurich in population, Geneva is 43 percent, and Bern is 39 percent (Kümmerley and Frey 1984, p. 23).

†Life in the Swiss community is not without its costs to the residents. If you are concerned with what your neighbors think of you, you are careful not to put out your garbage except on the days of pickup. You are careful to keep your house in good order and with an attractive exterior. You are careful not to let your children run wild, and if your children misbehave, you make an effort to apologize to their victims and to prevent a reoccurrence. You curb your dog. You support your family. You work to make your marriage a success, because among other reasons, a divorce is difficult to explain to your neighbors. You even keep your car washed. (In one town one of the

SCHOOLING

Unlike the role of education as "a continuous questioning of the existing world," a view held by many liberal educators in the Western nations, Lüscher, Ritter and Gross (1973) found a basic conservative ideology at the base of Swiss education. Under the Swiss model, "education is seen as a means of integrating children into the 'system' and helping them to internalize the norms of the existing order" (p. 89).

Just as failure on the job for an adult is more serious in Switzerland, where work has priority over many other activities, so does failure in school become a matter of concern not only for the school personnel but also for the child's family, relatives, family friends, various agencies, and interested people in the community. Because of the concern for the family's Ruf ("family reputation" in the community), and because in Switzerland more than in many other countries completion of basic education is the path to economic success and self-sufficiency, parents and children both take the educational process seriously. Observers of Swiss education have referred to the heavy emphasis placed on scholarship and discipline in the schools, the high status of teachers, and the well-established role obligations and social distance traditionally expected in contacts between students and teachers. The Swiss school has few extracurricular activities and group programs, but it has a heavy academic load (Clinard 1978).

The integrating of children into the economic, social, and political system by Swiss schools includes emphasis on the critically important norms and values of the society. Among these values and norms, a key item is, again, the principle of subsidiarity.

Under this principle, according to Lüscher, Ritter, and Gross (1973, p. 15), "only those tasks which cannot be fulfilled on the initiative of the individual should be taken away from him and given to the family or community. The same is true of the relationship between smaller groups, such as the family, community, canton, and the federal state." Thus, the family is primary, the community has priority after that, and only then can the canton or federal state make a claim on the loyalty of the individual. Similarly, private and voluntary institutions have priority over state institutions. Thus, loyalty to the family and community is instilled early in schoolchildren.

---

residents left his dirty car in front of his house for a few weeks, only to find his neighbors washing it for him one Saturday morning because "they were trying to be helpful to him.") Of course, you also avoid deviance in action or behavior, and if you do deviate, you are prepared to explain your reasons to your neighbors and friends.

Along with this is a concern for the effect of one's behavior on the rights of others, based not only on the Golden Rule but also on the concern for what others will think (Clinard 1978). The norms are taught to schoolchildren, including adherence to local authority, to the consensus, and to amicable agreement and the responsibility of the individual to the group. These are also enforced in the schools. Traditional models of behavior, such as respect for and adherence to the wishes of elders, teachers, and parents, are all required of schoolchildren. Unlike the "Lonely Crowd" model of Reisman (1950), however, children are taught to carry individual responsibility for their actions, whether they occur in groups or alone (Clinard 1978). Finally, children have instilled within them the importance of doing their work and of seeking its completion at a level of the highest quality. Gruner (1982) has indicated that from early childhood to old age, "the true Swiss is ready and willing to work, rising early and working late" as if for the joy of it. During interviews with school officials, the same theme was repeated again and again: Switzerland has no real resources except for its people. People can have value to the society only to the degree that they are competent, trained, and eager to do their best work. Finally, children are taught to rely only on themselves. This adds up to a lifelong theme of economic self-reliance, postponed gratification, and conservation of assets. Thus, as was noted, Swiss citizens have one of the highest rates in the world of bank savings and one of the highest rates of voluntary health and other insurances (almost 95 percent).

Closely tied to the values and norms taught to children at home and in school is yet another norm of behavior. This has been described as "the norm of reciprocity" by U.S. sociologists. This norm is based on exchange theory concepts and essentially indicates that there is "no free lunch" in any society. This means that people cannot get "something for nothing" and that if someone secures something without effort, then someone else must have made an effort for that person. Thus, Swiss children are taught at home and in school that if they are to expect goods and services, they must, in turn, do their work (learning) well and carry out the chores assigned to them. Children soon learn that such a quid pro quo arrangement amounts to an ethic of not relying on others to do your share of the work. This ethic is reinforced in school, in the community, on the job, in the workplace, and in the Swiss army, which embraces almost all adult able-bodied males. The Swiss child, unlike others, is answerable to the family for all actions and inactions in school and in the community (Egger 1978).

The status of teachers in the Swiss community is also a critical element in terms of effective learning of the students. Teachers in Switzerland are paid better than in many Western countries. An elementary school teacher may begin at Fr40,000 to Fr50,000 and can earn up to Fr70,000 in time. The Swiss high school teacher earns up to Fr80,000 to Fr100,000. The high salaries reflect a much higher level of public respect for teachers. Thus, a teacher's recommendation for a child's learning activity is usually accepted by the child's parents. Many schools do not have principals. Instead, the school is something like a community of teachers who operate the school under the policies of the local elected officials. In some localities, the teachers are voted in by the electorate. In many localities, teachers provide strong political leadership. Thus, many teachers are motivated to work with the parents and citizens of the community. Parents are considered by teachers as a crucial element in the learning equation of children, and this accounts, in great part, for the high success rating of Swiss education.

Research has shown that parents are a critical factor in the academic and vocational achievement of their children. Stevenson (1983) conducted comparative studies of children in Japan, Taiwan, and the United States and found that academic success occurs primarily when parent-child interaction takes place that involves the substance of the learning tasks. By parental helping, encouraging, and personal involvement, children make academic progress even when other factors may be weak or missing. Stevenson's study indicates that when there was parental concern with the child's academic progress, the child spent less time playing and watching television and more time doing homework and reading. Similarly, the children of better-educated parents achieved more in the academic realm. Where the parents were involved in the child's learning, the teachers made more progress with the child's learning goals. With a high level of literacy and education as well as a high level of respect for learning among Swiss parents, and with a high rate of involvement in the learning tasks of their children, fewer Swiss children fail to make progress in academic and occupational pursuits. This progress ensures the child's avoidance of a life of welfarization.

In Switzerland, if competent in school and career training, the rewards are apprenticeships, occupational acceptance, and a place in the adult community.* The Swiss individual is a self-reliant, work-

---

*The Swiss design of occupational training derives from the original guild programs for preparation of a master of a trade. This system, unlike the English and U.S. form based on specialization and division of labor, ensures that every student learns to become a generalized master. Because of this, Swiss industries are primarily small—often a total of nine workers for one product. This requires much more training and competence but ensures continued employ-

addicted, proud member of the community, with a secure place in a society that Clinard (1978) observes "has never had the marked disparities in distribution of wealth that characterizes many other European countries. The position of the Swiss worker today is generally more favorable than that of workers in most other countries" (p. 112). Such an individual could hardly become chronically dependent upon public aid. Unlike other nations, Switzerland offers no exciting inducements for those who do not complete schooling and career training. The path to rewards is clear.

THE COMMUNITY YOUTH AUTHORITY

Still another pattern of welfare dependency prevention operates out of the community youth authorities in each town. The Youth Authority is a publicly supported recreation, social service, supervision, and career program that serves a high proportion of the youth in each town. In no other country are such extensive and intensive efforts made to aid and direct a nation's youth into social activities and away from self-destructive and wasteful "sidetracks." The degree and amount of services given to young people and the attempt to tie recreation to character and career development among youth are both effective and unusual. There is a close working relationship between public welfare services and the Youth Authority that ensures that families given financial aid are also provided with supervision of their parenting methods when this is appropriate. The ease with which Beistand ("supervision by locally appointed guardians") is established and the local emphasis that such services shall include Rat und Tat ("advice and action") is probably not matched elsewhere. Where no Youth Authority is available (in smaller towns and villages), the responsibility of youth work is carried out by church agencies, which are often provided with official status.

EMPLOYMENT

In Switzerland more than in most societies, what an individual does for a living defines a person. How well the individual does it determines the degree of respect received from neighbors and community. The importance given to career preparation, vocational training, and professional education in the Swiss society has been in-

---

ability for those who complete the apprenticeship. Most workers are both interchangeable and adaptable within one industry.

dicated. Because work is so highly valued in the society, the individual worker usually seeks to perform well on the job. Despite the high level of employment in Switzerland, the loss of a valued job can have a devastating effect on a person. The turnover rate for employment in Switzerland is relatively low, and many employees tend to remain with one firm until retirement. More so than in many other countries, Swiss workers gain social as well as economic benefits from their jobs. Thus, the employment setting serves to prevent people from seeking public dependency not only by providing them with income but also by influencing their behavior and performance both on and off the job. In a nation with less than 1 percent unemployment, and in which opportunities abound for the qualified, the Swiss scene rewards employment and devalues welfare dependency.

MILITARY SERVICE

As indicated previously, all adult males in Switzerland are subject to compulsory military service until age 50. Every young adult who is able-bodied is required to spend some months in basic training, followed by annual service in a unit near his home. At age 20, men undergo a 17-week basic training. For the next 12 years, training continues for approximately two weeks per year. As a result of this, there is a sense of mutual camaraderie reflected in virtually every sector of Swiss society (McPhee 1982; Taki 1983). Employers are required to fund the wages of employees while on military duty, and many companies are glad to do so because military contacts are often good for business.

If the norms and values of the Swiss society and the loyalties of the locality have not been fully absorbed by young Swiss males, military service with a local unit will usually complete the process. Each soldier in the Swiss army is required to operate with a group and is also called upon to do his share within the group and on individual assignments. Thus, Swiss army experience serves to reinforce those qualities that prevent and operate counter to the welfarization process. An individual who "goofs off" at the expense of his unit has to cope with criticism not only during the annual encampment but also in his home community where other men know him. Thus, to request welfare without adequate reason or to not support his wife (or ex-wife) and children would serve to stigmatize a man in his community. Because a person's reputation in the community is critically important, it is carefully guarded in terms of appropriate behavior in the community. This is not conducive to welfare dependency.

Schweizer (1985) believes that the military is only one of a number of cohesive factors in the extensive Swiss social system. As in

other countries, those who are like-minded and equally desirous of
finding solutions to mutual problems or in pursuing joint goals find
themselves together in organizations, guilds, and militia units. The
societal system is such that many of the problems of the community
and society are informally solved in group sessions. Just as many
U.S. corporate problems are solved on the golf courses, many Swiss
community issues are discussed during the men's free time in military service. The difference, however, is that the U.S. corporate
golf course circle is hardly as democratic as the Swiss community
militia unit. Similarly, during free time at the local voluntary fire
and water departments, community men and women who come together will discuss problems of community concern. There are many
other joint activities in the communities, including the voluntary ava-
lanche protection service, the snow and rescue service, and the sani-
tary (first aid and volunteer ambulance) service, where community
issues are frequently discussed. (Only the major towns have professional fire departments.) The various social (formerly vocational)
guilds, in which most Swiss have at least one membership—and many
have as many as five memberships—also serve the common purposes.
Many of these guilds have idealistic goals that are appropriate to cohesive community efforts. There are also sport guilds, women's
guilds, animal breeder guilds, guilds to serve the blind, the Red
Cross, and the Blue Cross guild, which promotes prevention and treatment of alcoholism. Some have only a few members, but many have
thousands of members (Hoffmann-Novotny et al. 1978). Almost one-
third of the population are to be found in more than one guild or organization. In Switzerland and in other parts of Europe, service clubs,
such as Rotary, are exclusive and focus on community leadership,
but most of the guilds and clubs contain a cross-section of the Swiss
population and classes. Another organization is the civil defense for
protection against wars and catastrophe. This is a large program,
including over five hundred and twenty thousand people. This organization is channeled to local community levels and reaches all levels.
The members have frequent local practice and demonstration programs
in cooperation with other nearby communities, organizations, and
military units. Much social interaction occurs during these programs,
and the closeness that men develop in their local military units continues on into civil defense, which involves men ages 50 to 55 and
others who are exempted from military service. Additionally, about
twenty thousand women are also involved in civil defense. This kind
of active involvement in the community at so many levels naturally
creates an atmosphere of cooperation and consensus. This kind of
cooperation and cohesion serves to promote self-sufficiency and community responsibility. Obviously, this kind of atmosphere also serves
to prevent and constrain welfarization.

SWISS DOMESTIC RELATIONS COURTS

Still another informal or preventive control of welfarization exists in the Swiss divorce courts. When one considers that welfarized poverty has become feminized in many Western nations as an aftermath of divorce and unmarried motherhood, it is noteworthy that these problems are relatively small in Switzerland. One reason can probably be found in the policies and effectiveness of Swiss divorce and family relations courts. There are rarely complaints about courts. On the contrary, there is considerable respect among welfare administrators for the helpfulness of the courts in not contributing to the growth of welfare. Apparently, husbands who are divorced from their families must provide sizable amounts for child and spousal support, and those who do not have their wages garnisheed. Funds are advanced automatically in many cantons to divorced wives with children by the welfare agency, which then secures reimbursement from the fathers by legal mechanisms. The fact that most fathers do pay what has been specified by the courts makes it more probable that they will also visit their children and thereby serve as a parental influence on them. The only exception to the fulfillment of paternal responsibilities occurs among those few families where the father is an absent foreigner and therefore out of the reach of the courts. The Swiss rate of paternal determination and paternal support (outside the marriage) is also much higher than in most other nations. The low birthrate by unmarried women is occasioned probably by the high levels of informal social control, and this has contributed to the pattern of low welfarization rates. If the father is not available, then the community requires the woman or her parents and siblings to provide support.

District family relations courts thus have a direct effect on the prevention of feminine welfarization. What is less apparent is the informal effect of such court processes. In a conversation with a Swiss banker, it was learned that he had provided a bond to one of his clients to guarantee child support. This happened when a divorced man from a small town was summoned to meet with a judge. The judge was concerned that the man's intended marriage to a second wife might result in his nonpayment of scheduled child and spousal supports. The judge indicated that although he was not legally concerned as a judge, he was concerned as a citizen of the community. He did not want to see the community having to support the man's first family. He indicated that because of his concern, he planned to appear at the impending wedding ceremony to protest the marriage as fundamentally immoral, in that the man intended to develop a new family at the community's expense. The judge was dissuaded from his plan to publicly protest the marriage only after the man had posted a bond, which his brothers, sisters, and parents participated in, and which guaranteed continued child and spousal support until the children were grown.

Most judges probably do not involve themselves to this extent in the enforcement of their decrees, but the fact that this does occasionally occur provides an indication of the informal community controls that serve to limit or prevent welfarization.

The job of the domestic relations courts is eased in many ways by the operation and structure of the Swiss society. Similarly, the operation of the court system is such that collection of child and spousal support can be enforced by the withholding of wages, by the establishment of liens on property, and by a variety of informal and social constraints. It must also be noted that, unlike the practice in many other countries, in Switzerland divorce is not always granted by judges. There is no no-fault divorce. Divorces are more difficult to secure, and people try harder to make their marriages work. Thus, the divorce court system also serves to limit the development and perpetuation of welfarization in Switzerland.

In consideration of the divorce process in Switzerland, it is helpful to note the observations of Stephen Jaffee (1977) on developments relating to the Southern California life-style, which tend to portray the unfolding future of U.S. attitudes, values, and norms. Jaffee reports that the contemporary U.S. life-style, as revealed particularly in Los Angeles, is antagonistic toward and destructive to sustained affectional relationships between men and women. He came up with five factors that contribute to the breakdown of such relationships.

The first is the American "revolution of unfulfilled expectations." Under this revolution, "People have come to expect more from their romantic adventures, just as they expect more from their kitchen appliances and their automobiles" (p. 7). These raised expectations of romantic rewards are in part the product of the host of "pop" and other psychotherapy movements that elevate the gratification expectations of the individual, often at the expense of family, work, and social group. This leads to a low threshold of disappointment. When a relationship problem arises, one or both of the individuals involved fall back on the ritual dialogue of their brand of self-improvement or assertion. When this does not work (and it is bound not to, since both assume that personal hedonistic gratification rather than mutual consensus is their goal), the family is torn asunder.

A second reason provided by Jaffee is the growing unpopularity in the United States of commitment, especially as it relates to romance and marriage. Commitment implies an understanding that both partners will work out problems as they arise. Without commitment, marriages are broken early or never undertaken. Without commitment, conciliation and adjustments are viewed as "giving in," weaknesses, and self-diminishing.

The third factor listed by Jaffee is the virtual destruction of the stigma of divorce. In the past, the norm of society was a long uninterrupted marriage state. Those who did not marry were viewed as deviant. Under this pattern, the fear of social and business ostracism caused partners to think carefully before testing the waters of divorce. But in present times, those still married are now viewed as deviants, and the sheer numbers of those separated and divorced lessens the fear of being singled out as a maladjusted or unfit person. (In Los Angeles County, divorces now outpace marriages.)

Jaffee lists the fourth factor in the destruction of the familial bind as the growth of the cult of narcissistic autonomy worship, where marriage and even responsible parenting are sacrificed on the altar of personal independence. He cites the booming business in courses, books, fads, and organizations that promote "how to look out for number one." Despite the view of more mature and less biased psychotherapists who indicate that isolated self-sufficiency (often at government expense) can be personally devastating in the long run, this push toward individualism and hyperautonomy is growing at the expense of effective socialization for children and at the expense of familial mutual aid and lasting relationships necessary in sickness, in misfortune, and in later life.

The fifth factor listed by Jaffee is the adoption of the no-fault divorce law in the United States. This law not only makes divorce easier to obtain by either party, but it deprives the participants of the opportunity to learn what went wrong in their relationship. The same type of failures in relationships keep repeating themselves with new partners for many of the participants. Jaffee also indicates that no-fault divorce provides no opportunity for resolution of the mutual anger between the participants. This well of anger often leads to a continuing war over custody of the children, division of property, visitation rights, and child support payments, and casts a severe toll on the well-being of the children and divorced parents.

It should be noted that the culture of aloneness and divorce in southern California is considered by many informed scholars as an indicator of continuing trends in the United States and even in some other Western nations. It should also be recognized that the "revolution of unfulfilled expectations" has already occurred in the U.S. underclass and in part of the populations of other welfare state nations, as reported in Chapter 1. Similarly, the U.S. underclass (Auletta 1982; Sharff 1981; Sheehan 1976) already suffers from a lack of commitment to career, family, and other activities that constrain their life alternatives and that, in reality, define their condition. The stigma of divorce (and desertion) has long been attenuated and diluted in the U.S. underclass. To a lesser extent, these underclass conditions have already appeared in other classes and nations.

Unlike the United States and other Western nations, in Switzerland there has been little evidence of a revolution of unfulfilled expec-

tations. The average Swiss knows very well that what one gets is
exactly what one has earned. Nothing more is expected, and thus,
there is seldom disappointment. Commitment is perhaps the name-
plate of the Swiss character. The child chooses a career and works
to achieve it. The Swiss undertake business enterprises and make
them function. Swiss workers undertake jobs and perform them to
their own and their employers' satisfaction. The Swiss people undertake residency in a community and, by and large, fulfill their responsibilities to that community. The stigma of divorce (and the unmarried state) prevails. The cult of the self has apparently not taken
root, perhaps because of the views on interdependency, consensus,
and subsidiarity. Finally, the concept of no-fault divorce has never
been accepted under Swiss law. Thus, unlike the unfolding U.S. (and
other "modern" national cultures), the Swiss culture supports the
institution of the family for all of its population, including those persons who come into contact with the Swiss public welfare programs.
The Swiss culture is clearly not supportive of family dissolution or
welfarization.

PSYCHIATRIC FACILITIES

     In addition to the community welfare offices that serve in a
multiservice capacity; the various other community agencies, such
as the youth authorities and the voluntary agencies (like the Pro-
Juventute, Pro-Senectute, and Pro-Infirmis); the various voluntary
social agencies, such as the Institutes for Family and Marriage; and
the variety of church agencies and services, all of which support
family functioning—in addition to these, there is also an extensive
system of outpatient and inpatient psychiatric services in Switzerland.
May's survey of mental health services in Europe (1976, p. 7, Table 1)
indicates that unlike most other nations, which designate a ministry
of health or an equivalent such as the Ministry of Social Affairs and
Public Health (as in the Netherlands) or the Ministry of Health and
Social Security (the United Kingdom), in Switzerland the responsibility for provision of mental health services rests with the cantonal
and local health authorities. In addition, there are nine voluntary
mental health associations in Switzerland (May 1976, p. 11, Table 3),
seven of which are concerned with mental health services and two with
mental retardation services on a national basis. Switzerland has 62
mental health outpatient facilities, an average of 2.4 per canton (p.
20, Table 7). The mental health outpatient facilities per 100,000
population is calculated at an index of 1.0 (based on the World Health
Organization formula), which compares favorably with countries such
as Belgium (0.8) and Luxembourg (0.6) and which is in the same ser-

vice range as France (1. 9) and the Netherlands (1. 7). Switzerland does not provide as many facilities as Sweden (4. 2) and Denmark (2. 4), where government services have displaced many of the self-help programs formerly dealt with by the family. In terms of inpatient psychiatric facilities, Switzerland's index is calculated at 2. 7 (p. 29, Table 11), which is comparable to Belgium (2. 8), Denmark (2. 5), France (2. 3), Luxembourg (3. 0), the Netherlands (2. 9), Norway (2. 6), Sweden (3. 5), and England and Wales (2. 8). In these institutions, Switzerland has fewer beds per psychiatrist than other Western nations (p. 37). Similarly, Switzerland has less admissions per psychiatrist than other Western nations listed.

It is apparent that Swiss mental health facilities are comparatively available, but it is also obvious that they are not so profuse and ubiquitous as to become an alternative to family self-care and services at the local and community level.

FAMILY POLICY

In addition to the active direct-service programs of support to individuals and families, there is also a national organization with substantial membership and involvement in each of the cantons and in most of the communities. This organization, Pro Familia, had its roots in a national conference on family life that took place in 1931 (Lüscher 1982). Subsequent conferences led to the establishment of the organization as a continuing entity shortly after World War II. Since then the organization has sponsored research on family life in Switzerland and has pressed for legislative support at all levels for programs and policies in support of cohesive family life. The organization provides annual conferences on family life. Among its accomplishments is the comprehensive work on family policy in Switzerland. This study was conducted by the Swiss Ministry of the Interior on the urging of Pro Familia. This work provides not only a tabulation and explanation of current Swiss family policy but also a set of recommendations and underlying rationales for suggested additions and changes in policy. Probably no other nation has produced such a comprehensive analysis of the strengths and weaknesses of its family policies (Switzerland, Ministry of the Interior 1982). Many nations, including the United States, actually have no family policy to guide legislative and leadership activity, with the result that much social legislation does not take into consideration the effect of such actions on family life. In many instances in other lands, such legislation often serves to weaken family life, but in Switzerland this is less likely with an active profamily movement and a ready documentation of established policies and recommendations.

The presence of an active movement in support of effective family life and its socialization processes serves as an additional constraint on welfarization.

SWISS WELFARIZATION

Presented here have been the formal and informal constraints on the welfarization process in Switzerland. It is clear that these constraints are effective in limiting welfare dependency. It is important to note that welfare experience over decades indicates that there is a functional juxtaposition between making welfare adequately available for those unable to help themselves yet controlling welfare so that it does not grow unnecessarily in numbers and time and so that it does not interfere with the employment marketplace. To constrain welfare too much by formal and informal methods may create a hardship for people who have no alternative but to seek it; but to make it too freely available and with too few controls will lead to its spread in the population over time. Excess availability with few controls will create dysfunctionality in the employment market along with other unhappy effects in relation to inflation, immigration, and a variety of social ills.

What about the Swiss experience in this regard? Have they been too harsh or too permissive? An examination of the estimated dependency caseload may help to answer the question.

In a search for transgenerational dependency undertaken by the author, evidence of dependent welfare in each of the public agencies in the four major localities and one smaller industrial town was sought. In each of these cities and towns, a major part of the welfare load is made up of what would be described in the United States as the "medically needy." These are people who have sufficient income, either from current earnings or from assets (such as earned social-insurance benefits), to be able to meet their basic needs. They include many aged, handicapped, and infirm people. Because there is 95 percent medical coverage in Switzerland and because there are some duplications of medical insurance, conservative estimates of those adequately covered for medical care ranges from 78 percent to 90 percent. The remainder are provided with medical coverage and reimbursed medical care as needed by their local public welfare agencies. This makes up a significant amount of the work of the Swiss public welfare agencies, ranging from 20 to 30 percent of the caseload.

Because there is no central agency to collect information and data on welfare caseload, and because services are only available in each locality, there is no generally uniform statistical format collected in each of the agencies. Thus, whatever data was available was supplemented with estimates from social workers and public welfare officials.

Data were not gathered on the temporarily dependent, primarily people who had used up their unemployment compensation or whose unemployment compensation was less than sufficient to meet immediate survival needs. This group also included a sizable proportion of people who were employed on a part-time basis. In almost all of these cases, the families were intact and were seriously searching for additional employment. Most of the unemployed or partially employed were in geographical areas of industrial recession and were discussing with their social workers possible occupational retraining programs or moving their families to areas where appropriate employment was becoming available. Many of these families were on the rolls only for continuation of payment of their interim medical insurance and were otherwise self-sufficient.

In all of the agencies visisted, and as reported in Chapter 5, there was a small part of their caseload (aside from the temporary clients, aged, and handicapped or medically needy), amounting to 5 to 10 percent, who represented dependent cases of long duration. These were described in terms that one would list as multiproblem families. In each instance, there were such factors as inadequate employment conditioning and/or training, psychiatric disturbance, a disorganized family life, disorganized handling of family funds, antisocial or incompetent role models for the children, divorced parents, parental neglect of children, distancing of the individuals from their extended families, excessive gambling, and so forth. In the analysis of these cases, it appeared that their number was few in proportion to the total welfare load and that measures were being taken by the public welfare agency either to resolve the dependency in the present generation or to prevent it in the next one. Great care was being taken to ensure that children in such multiproblem families were being carefully observed and protected from the destructive life patterns of their parents. In some instances, aid for such families was being maintained only as long as the parents accepted supervision and guidance of their childrearing patterns. In other instances, Beistand ("guardianship" or "supervision") actions were taken in cooperation with the Youth Authority to ensure that parental responsibilities were fulfilled. In still other instances, children were placed in special homes and institutions, where reportedly more than a majority succeeded in maturing into responsible productive citizens.

In discussing this portion of the public welfare caseload in the agencies visited, it was learned that the use of placement of children in institutions is very much on the decrease. Instead, families with children are currently being given a combination of aid, advice, and supervision to ensure that the children are appropriately reared to fit into an employed, self-sufficient population.

In attempting to find the existence of transgenerational poverty, representatives of the ATD Vierte Welt (the "Fourth World") were sought out and interviewed, including the author of a book on poverty in Switzerland. They estimate that 3 to 5 percent of the entire Swiss population are transgenerationally dependent, based on their belief that as unemployment rises, more people fall into structurally caused poverty and remain there because of an inability to again rise from the aided-population level. This seems questionable, since at most, the Swiss unemployment rate is less than 1 percent, and this includes many of the partially employed population.

Many of the unemployed are individuals in a family where two or more persons are employed. Thus, it is hard to believe that few of the unemployed ever become self-sufficient again. It was learned that the ATD Vierte Welt organization had contact with some 2,500 Swiss families whom they have hosted in their country-farm vacation program over the past 15 years. They believe that these families are continually in need. Both partners in most of these families, they believe, were themselves brought up in poor families. In each of these families, at least one of the partners has been, or is, under Vormundschaft ("legal supervision") after a period of alcoholism or psychiatric care. Many of the children in these families have been placed in children's institutions or foster homes, usually against the wishes of the parents, based on the reported inability of the parents to provide an appropriate rearing of the children. These families reportedly are unable to find adequately remunerative employment and to help their children to adequately utilize the school and apprenticeship system to become self-sufficient when they reach adulthood. Many of these families have more children than they know how to deal with. The ability of these families to deal with the realities of everyday life is only marginal. Some are ex-offenders. The individual problems of the families keep repeating themselves because these families find it impossible to avoid recurring problems or to learn from previous difficulties. These families have constant problems of relating to community institutions. Their relations generally are in turmoil. All have problems of Ruf ("reputation"), which make employment, housing, and so forth difficult. From the perceptions of these clients, the community and its institutions seem threatening. Many are the victims (and causes) of accumulated unpaid debts and installment payments for items purchased and no longer used.

The ATD Vierte Welt portrayal of these problem families fits the description of A Welfare Mother described by Sheehan and The Underclass described by Auletta in the U.S. welfare scene, as well as Forman's (1982) description of dependent mothers and children.

Although claims of a larger recurring dependent population were made by the ATD Vierte Welt representatives, it became apparent the 15-year listing of 2,500 families might be a fair estimate of the number of such families in all of Switzerland. In interviews with the major public welfare agencies, it became clear that less than 3,000 families in Switzerland are multiproblem families.

These multiproblem families can be designated "individual poverty" rather than "welfare dependent" in terms of the Strang typology, based on the fact that there has been little politization of the welfare issue in Switzerland, unlike most countries, where the welfare clientele and the welfare administration have become symbiotically perpetuative. Similarly, there is almost no "bending" of Swiss welfare policies or administrative procedures and a rather constant and firm adherence to the welfare-agency goals of rehabilitation.

In a population of approximately 6,500,000, a residual welfare dependency figure of even 3,000 represents less than 0.01 percent (less than one-tenth of 1 percent). This compares favorably with the estimates for many other nations, as noted in Chapter 1. Thus, it is believed that the Swiss level of chronic dependence is far less than that in other Western nations.

COUNTER FACTORS

There are a number of other factors, however, that may serve to weaken Switzerland's prevention of prolonged dependency. These include the loosening of family controls as exhibited by youth unrest, alcohol and drug abuse, and the effects of Swiss informal social control.

Alcoholism has long been a problem in Switzerland, and it is probably explainable in the light of Swiss social control. For decades each of the cantons has expended efforts relating to this problem. A special voluntary agency was established to promote nonalocholic beverages and to educate the public against use of alcohol. The special drug subcommission of the federal government estimates that about 10.5 liters are ingested annually per resident (somewhat over two gallons). This indicates that Switzerland is about tenth in Europe in the overall ingestion of alcohol. The ingestion of alcohol is concentrated in the male population of ages 15 to 75. Hoffmann-Novotny et al. (1978) indicates that about 25 percent of the population are heavy drinkers, with the highest ingestion between the ages of 25 to 54 years (p. 80). The institute found that the alcohol problem surpasses the drug abuse problem. The institute did find that many of the heavy abusers of alcohol and drugs at age 19 learn to abstain or to become light or middle-range consumers of alcohol by age 22. Only 26 percent of the heavy users of alcohol remain heavy users by age 22 (pp. 81-83). Similarly, only 23 percent of the drug abusers remain abusers by that age. It is estimated by Reist and Wagner (1980) that between 130,000 and 140,000 people are serious alcoholics in Switzerland. This is about 2 percent of the population. Clinard (1978) also deals with the problem of drug addiction in Switzerland. He reports that 92.2 percent of the drug cases involved males under age 25. Clinard views drug use as a form of protest by some Swiss youth against the more conservative society, much in the same way that some youth will use unusual or bizarre hair and dress styles to indicate their differences with their elders. Trimborn (1982) reported that 107 drug-related deaths occurred in Switzerland in 1981, which represents about 1.7 such deaths per 100,000 people. In the United States, the rate is closer to 4.0 per 100,000 people.

Müller (1981) relates Swiss alcoholism differentially to the three major cultures of Europe. He indicates that the Italian drinking culture assesses alcoholics in a more tolerant way than the French and German drinking cultures. This tolerance of alcoholism implies less negative sanctioning of deviant drinking. The label "wine country," attributed to Switzerland, is mainly due to the fact that Latin minorities drink large quantities of wine. The differential ingestion of wine, beer, and hard liquors in the differing cultural areas of Switzerland supports Müller's findings. (Beer drinking is heaviest in the German-speaking area, and wine is more heavily used in the French and Italian areas.) Müller (1983) reports that alcohol is primarily a male drug in Switzerland, as reflected in the comparison of male and female mortality rates for cirrhosis of the liver. He reports that death because of alcohol intoxication is very rare in Switzerland. This is in contrast to the findings in the northern and eastern countries of Europe. In Switzerland he finds that alcohol is seldom used to achieve a condition of heavy intoxication. Müller indicates that alcoholic beverages have the function of situation markers that structure everyday life. Death because of alcohol poisoning in Switzerland is more likely related to accidents rather than to (intended or unintended) suicide. Acute alcohol intoxication is not a principal cause of death for the young but may be so for the middle-aged. Only 1.5 to 2.5 percent of all deaths in Switzerland could be traced to alcohol psychoses, and this has been decreasing since the 1960s as a result of improved treatment methods. Alcohol-related traffic accidents have increased in relation to total accidents since the early 1960s, from 6.8 percent in 1963 to 9.6 percent in 1978.

Comparative alcoholic consumption rates for Switzerland and other Western nations are reported by the Swiss Federal Office of Social Security (1978, pp. 114-15). The use of wine, beer, heavy spirits, and total intake are shown in Table 5.3.

Thus, it appears that there is an average of 9.9 liters of total alcohol consumption in all 11 countries listed in the table. Switzerland is 0.2 liters above the norm and is fourth in the consumption of alcohol per person, with six nations having heavier rates of consumption per person.

TABLE 6.1

Comparison of Alcohol Consumption across Countries

(annual liters per person)

	Wine	Beer	Heavy Spirits	Total
Switzerland (1975)	44	72	4.8	10.1
France (1975)	104	45	6.5	17.0
Italy (1975)	107	13	5.0	13.4
West Germany (1975)	23	148	7.6	12.5
Austria (1975)	35	104	4.1	11.1
England (1974)	5	110	2.5	7.4
Netherlands (1974)	8	69	3.5	8.8
Denmark (1974)	8	120	3.5	8.9
Norway (1974)	3	41	5.4	3.9
Sweden (1974)	7	57	5.4	6.2
United States (1974)	6	75	6.0	10.1

Source: Switzerland, Federal Office of Social Security, Bericht über die Lage der Familie in der Schweiz [Report of the position of the family in Switzerland] (Bern: Federal Office of Social Security, 1978), pp. 114–15.

177

Reist and Wagner's 1980 report estimates 13,000 to 15,000 drug abusers in Switzerland. Clinard (1978) believes that "although these figures may alarm the Swiss, they show far less drug use than in the United States and many other countries" (p. 45). For example, in a sample of over four thousand Swiss army recruits at age 22, 77 percent in 1975 were found to have never used drugs. Follow-up studies on these young men during military service showed little change in the abstinence rate. Victor Reidi of the Bern Youth Authority (see Trimborn 1982) also believes that drug abuse among a sector of Swiss youth is a form of protest against Swiss conformity and performance requirements. Those who succeed in meeting the society's requirements and standards are less likely to become abusers. In a society such as the United States, nonconformity and less-adequate competency would be more readily tolerated. For example, chronic drug and alcohol abusers who are young adults, and who are thereby considered totally disabled, receive a monthly grant in the Supplementary Security Income program in the United States. (In California the grant is $440 per month, tax free, plus medical care.) It is as if, in the United States, society, the state, and the federal government had legitimated drug abuse. Reidi views the Swiss situation as more demanding of youth. "I think there is no country in the world with more rules, more intolerance. Everybody in Berne is a village policeman, and in Zurich it is even worse" (Trimborn 1982, p. 1). What Riedi probably does not realize is that if there was less social control in Bern, the rate of youth deviance and drug addiction would be more serious.

According to Trimborn, the Federal Health Office estimated that there were 6,000 heroin addicts in Switzerland in 1981, and he believed that it has grown by 2,000 per year since then. Even with such growth, the drug rate per 10,000 population is still minimal compared with the rates in other Western nations. Drug-related arrests in Switzerland, according to Trimborn, were 1,518 in 1980 and 9,699 in 1981. Convictions, however, were 5,581 in 1980 and 6,460 in 1981. Of the 1981 drug arrests, about 1,900 were foreign residents. Unlike the situation in other countries, the major traffickers of hard drugs in Switzerland, according to Trimborn, are foreign residents—primarily from Pakistan, Turkey, and Italy.

Clinard (1978) indicates that the pattern of Swiss drug-use rationale by youth is different from that in other societies. In other countries, drug use by youth is often tied to crime, including violent offenses. The drug addict will steal, rob, or assault others in order to satisfy his or her habit. Much of the U.S. drug culture is also tied to crime and delinquency. Swiss youth, according to Clinard, use drugs primarily as protest, but they will "generally not steal the property of another as this would harm someone, [but] they believe that their use of drugs is a personal decision" (p. 132). Thus, the Swiss drug problem is less tied to the drug-use patterns experienced elsewhere in the world, where the drug addict has few or no internalized social controls.

In any case, the alcohol and drug abuse problems in Switzerland, while relatively light in comparison with some other Western nations, is still a serious concern to Swiss policy makers. Although the Swiss external controls are probably more effective in detecting and reaching the users and in seeking some treatment and rehabilitation for those affected, it should be noted that the problem is a growing one. The Swiss drug problem is compounded by contagion with visitors and cultures from other countries where drug abuse is more common.

Still another problem has concerned the Swiss in recent years. This relates to actions taken by Swiss youth in the early 1980s to express their objection to the way in which the adult world dealt with them. The protest primarily arose in Zurich, but it was also evident in other Swiss towns. In Zurich the protests began in 1980 in connection with plans to rebuild and refurbish the community opera house. The principal complaint of the youth was that the opera house was for adults, and they wanted the community to provide them with a youth center. Unfortunately, the police reacted to the protests with force rather than with flexible negotiation. After considerable damage to expensive show windows and some degree of rioting, which would have been less-seriously considered in other nations, a youth center was established. For a time, the youth demanded self-supervision in the center, but this turned out to be a "cover" for the sale and distribution of illicit drugs. Finally, an arrangement was made to provide center facilities under adult supervision. The newspapers of that time indicated that there had been considerable involvement by drug-culture youth from other countries who happened to visit Zurich at the time of the riots. Taki (1983) reports that "the Swiss even found evidence of KGB involvement in the Zurich youth riots of 1981" (p. 32). It also should be noted that despite the involvement of many Swiss young adults (to whom rifles and ammunition, which are kept at home, are regularly issued), not one young adult brought a weapon to the riot. Thus, it appears that the riot was intended to be peaceful and not destructive on the part of most of the youth.

As an aftermath of the riots, a special study was formed by the Swiss Federal Advisory Commission for Youth Affairs, and their preliminary report was published in November 1980 (see Fricker 1981). A set of theses on the 1980 youth disturbances was included in the report. These theses included the issues of "radical minorities who felt misunderstood by the adult community," alienated, aimless, "speechless," and having to resort to violence to secure adult attention (pp. 18-22). The causes of their unrest were laid to failures of the family to provide an atmosphere of warmth and emotional security and, primarily, to a sense that youth did not have a medium of communication with the adult community. Unlike the youth in other countries, who complain of lack of opportunity and career openings, the Swiss youth complaint is frequently that everything is set out for them "in concrete" and that too much is provided and predetermined for them. Unlike youth elsewhere, who complain about unemployment, Swiss youth complain about the reverse.

Considerable discussion about these issues and theses occurred in the years following the riots. Most towns now have youth centers with mutual management between staffs and youth. There has been work done in communities on the strengthening of youth–adult communication. The problem is also slowly being resolved by the aging of previously rebellious youth, who are now young adults entering decision–making circles.

Much of the youth unrest can be considered to have arisen from the lack of youthful opportunities for open deviance, opportunities available to adults in Switzerland through the many spring carnivals. Much of the youth unrest can also be considered to be a natural effect of adolescence. Still another cause may have been youth disturbances in other countries that contagiously reached Switzerland.

Swiss youth is probably under more pressure than youth in other countries, where career failure and incompetence are tolerated, condoned, and even subsidized by welfare and disability programs. Obviously, such unrest is to be expected, and the Swiss communities have begun to provide arrangements for dealing with these currents.

Swiss youth who have difficulty in conforming to the higher Swiss standards of behavior and competence are likely, as time goes on, to demonstrate their objections to the demands made on them and to wish for the alternative (but economically parasitic) life–styles available to youth elsewhere. Switzerland, however, provides hope for Swiss youth in terms of continuing opportunities.* Upward mobility is apparently still available in Switzerland—more so than in many other countries, including the welfare states. Berelson and Steiner (1964, p. 473) present a comparative index of upward and downward mobility in nonfarm populations by countries.†

---

*There are, unfortunately, no community or junior colleges in Switzerland, although interest in this form of alternative "catch-up" education has been discussed from time to time among some Swiss educators.

†We must note, however, that at the time of this study (1964), West Germany, France, and Switzerland were affected by the insertion of foreign workers into the social stratification patterns at the

Country	Upward Mobility (nonmanual sons of manual fathers)	Downward Mobility (manual sons of nonmanual fathers)
United States	33	26
West Germany	29	32
Sweden	31	24
Japan	36	22
France	39	20
Switzerland	45	13

Thus, Switzerland shows the highest upward-mobility rate and the lowest downward-mobility rate. If the measure of a welfare state is to help people develop to their fullest potential and to keep people from falling to lower levels of society, then Switzerland may be considered more of a welfare state than the United States, Sweden, France, Japan, or West Germany.

As long as Swiss society provides the opportunities for its youth to move upward, Swiss families hold together, Swiss schooling continues to influence youth, the community concerns itself with social problems, and apprenticeships are completed and jobs are secured, it is likely that Swiss youth will continue to conform and perform in the society. The same factors that tend to prevent welfarization also tend to constrain the alcohol and drug problems.

It should be noted that the 1980 and 1981 Swiss youth riots were quite unlike the 1968 riots, which involved primarily college students and the children of well-to-do families. In the 1980 and 1981 riots, Swiss youth were believed to be primarily youth with lower expectations, representing the small percentage of nongraduates of high schools and youths with apprenticeships of lesser status and rewards. If the Swiss communities are to do something for those youth (with lowered expectations who exhibit an anomic orientation) then increased "second-chance" educational facilities are needed. Such institutions, similar to the junior colleges on the U.S. scene, would make it possible for youth and young adults at the marginal apprentice levels to move upward and, perhaps, even to become eligible for college-level education. The increase of such second-chance opportunities for young adults in Switzerland is an important element in keeping the Swiss society permeable and democratic.

The difference between the problems of the Swiss agencies and those of other countries is that in Switzerland, almost the entire

---

lowest levels. These findings are therefore somewhat offset by these factors.

community seeks to prevent and control social problems. Unlike other nations, the Swiss society continues to have strong potentials for controlling and preventing welfare dependence.

REFERENCES

American Women's Club of Switzerland. 1982. "Social Shock." In Annual Report 1981/82, p. 4. Zurich: American Women's Club.

Auletta, Ken. 1982. The Underclass. New York: Random House.

Berelson, Bernard, and Gary A. Steiner. 1964. Human Behavior: An Inventory of Scientific Findings. New York: Harcourt, Brace & World.

Clinard, Marshall. 1978. Cities without Crime: The Case of Switerland. London: Cambridge University Press.

Egger, Eugene. 1978. "Education in Switzerland." In Modern Switzerland, edited by J. Murray Luck, Hugo Aebi, Joseph von Ah, Lukas Burckhardt, Erich Gruner, and Hans Haug, pp. 227-54. Palo Alto, Calif.: Society for the Promotion of Science and Scholarship.

Forman, Rachel Zinder. 1982. Let Us Now Priase Obscure Women. Washington, D.C.: University Press of America.

Fricker, Haus-Peter. 1981. Youth Disturbances in Switzerland. Zurich: Pro Helvetia.

Fuchs, Victor R. 1983. How We Live: An Economic Perspective on Americans from Birth to Death. Cambridge, Mass.: Harvard University Press.

Gruner, Erich. 1982. Private correspondence with the author. March 26.

Hoffmann-Novotny, Hans-Joachim. 1983. "Gesamtgesellschaftliche Aspekte der Entwicklung von Ehe, Familie und Fertilitat" [General community aspects of development relating to marriage, family, and fertility]. In Planspiel familie, edited by F. Höpflinger and Hans-Joachim Hoffmann-Novotny, pp. 1-26. Diessenhofen: Rüegger.

Hoffmann-Novotny, Hans-Joachim, Robert Blancpain, François Höpflinger, Martin Killias, Mathias Peters, and Peter Zeugin, eds. 1978. Almanach der Schweiz. Bern: Peter Lang.

Jaffe, Stephen R. 1977. "Marriage Is Alive but Not Well in Los Angeles." Privately published paper. (Obtainable from Mr. Jaffe's law firm in Los Angeles.)

Kubly, Herbert. 1981. Natives Return. New York: Stein & Day.

Kümmerly and Frey. 1984. Switzerland 1984: People, State, Economy, Culture. Bern: Kümmerly & Frey.

Luck, J. Murray. 1978. "Introduction." In Modern Switzerland, edited by J. Murray Luck, Hugo Aebi, Joseph von Ah, Lukas Burckhardt, Erich Gruner, and Hans Haug, pp. xi-xvi. Palo Alto, Calif.: Society for the Promotion of Science and Scholarship.

Lüscher, Kurt. 1983. "Die Schweizer Familien der achtziger Jahr" [The Swiss family in 1980]. Neue Zürcher Zeitung, October 10, pp. 18-19.

_____. 1982. "Fifty Years of Family Policy in Switzerland." In Pro Familia, report of the Family Conference of November 21, 1981. Lucerne: Pro Familia.

Lüscher, Kurt K., Verena Ritter, and Peter Gross. 1973. Early Child Care in Switzerland. London: Gordon & Breach.

McPhee, John. 1982. "A Reporter at Large: The Swiss Army." New Yorker, October 26, pp. 50-117; and November 2, pp. 55-112.

May, A. R. 1976. Mental Health Services in Europe: Review of Data Collected in Response to a Questionnaire. Geneva: World Health Organization.

Müller, Richard. 1983. "Alcohol Problems in Switzerland." In Consequences of Drinking, edited by N. Gieslesecht, Norman Giesbrecht, Monique Cahannes, Jacek Moskalewicz, J. Österbergiesa, and Robin Room, pp. 45-70. Toronto.

_____. 1981. "Contemporary Patterns of Drinking in the Cultures of Switzerland." Contemporary Drug Problems 10:155-77.

Newspaper Enterprise Association. 1985. World Almanac and Book of Facts, 1985. New York: Newspaper Enterprise.

Reist, W., and R. Wagner. 1980. "Zur Drogenproblematic in der Schweiz" [On the drug problems in Switzerland]. Zeitschrift für offentliche Fürsorge, February, pp. 162–66.

Riesman, David. 1950. The Lonely Crowd. New Haven, Conn.: Yale University Press.

Schmid, Carol L. 1981. Conflict and Consensus in Switzerland. Berkeley and Los Angeles: University of California Press.

Schweizer, Willy. 1985. Private correspondence with the author. January 11.

Sharff, Jagno Wojcicka. 1981. "Free Enterprise and the Ghetto Family." Psychology Today, vol. 15, no. 4 (March).

Sheehan, Susan. 1976. A Welfare Mother. New York: New American Library, Mentor.

Statistisches Jahrbuch der Schweiz, 1982 [Statistical yearbook of Switzerland, 1982]. 1982. Basel: Birkhauser.

Stevenson, Harold W. 1983. "Making the Grade: School Achievement in Japan, Taiwan, and the United States." Annual Report, 1983. Stanford, Calif.: Center for Advanced Study in Behavioral Science, pp. 41–51.

Swiss Almanac of the Zurich Sociological Institute. 1978. Bern: Peter Lang.

Switzerland, Federal Office of Social Security. 1978. Bericht über die Lage der Familie in der Schweiz [Report of the position of the family in Switzerland]. Bern: Federal Office of Social Security.

Switzerland, Ministry of the Interior. 1982. Familien Politik in der Schweiz. Report of the Chair of Federal Department of the Interior, Work Group on Family Policy. Bern: Federal Office of Social Security.

Taki, Theodoracopulos. 1983. "Armed Neutrality." American Spectator, October, p. 32.

Trimborn, Harry. 1982. "Switzerland Faces Up to Growing Problem with Hard Drugs." Los Angeles Times, December 15, pt. IB, p. 1.

U.S., Congress, House, Select Subcommittee on Children, Youth, and Families. 1983. U.S. Children and Their Families: Current Conditions and Recent Trends. 98th Cong., 1st sess., May.

7

CONCLUSION

LESSONS FOR THE WESTERN WORLD

This volume has examined the negative effects of welfare-state dynamics on the populations of the United States and other Western nations and presented the conditions in Switzerland that have achieved the humane goals of the welfare state without negative effects on the population. Now, we approach the critical question: How has the welfare state survived in the West in the face of its obvious shortcomings?

Anderson, Lait, and Marsland (1981) have examined this question. They found that the welfare state in most countries has established for itself an external image that belies the underlying mechanisms. Thus, the welfare state is perceived as normal and permanent, despite the fact that it is less than fifty years old in many countries that are many centuries old. The welfare state is perceived by the public as safe, while changes to the welfare state are often viewed as risky. The welfare state is generally perceived as supported overwhelmingly by the population, although actually, its support lies primarily in a host of welfare-state servants and a secondary welfare industry, including academics, journalists, publicists, professions, and unions that depend on it. Although the public perceives the evaluators of the welfare state as objective, the facts are that those who evaluate it are selected insiders who are less accountable to the society and heavily accountable to the welfare-state organs that select them. Although the public perceives the welfare state as open to reporting in the media, the facts are that many of the privately admitted truths never reach the public media or public hearings. Although the public perceives the welfare state as open to objective criticism, the facts are that a special vocabulary is produced by the welfare state for its protection. Anderson, Lait, and Marsland indicate that the world has been reordered by the welfare state in heroic fashion in which any criticism or question of the state is interpreted in the vein of the needy being attacked by the "elitists," the "unscrupulous," and other cardboard characters. The public is propagandized to perceive employees of the welfare state as "public servants" and the academics who support the welfare state as "objective analysts." Thus, the "welfarists manage to convince nonwelfarists that they have some sort of special claim to altruism" (p. 27). The public image of welfare-state studies is that, supposedly, they are the result of objective analysis, but the facts are that critics, or even neutral analysts, are denied access to information necessary for a truly objective analysis of the welfare state. Much welfare-state research that raises fundamental questions dies aborning, blocked by lack of funding and "professional" objections based on "confidentiality." Anderson and coauthors believe that the welfare state places the burden of proof unfairly on the critic and questioner, rather than on itself, to assure the public that its obligation to provide aid is carried out more effectively, more economically, and with less social costs than other systems that could be devised (pp. 23-31).

The public also finds many aspects of the welfare state attractive. The income-redistribution system of the welfare state generally relieves the middle class of any concern it may have for the poor. Instead of having to become involved in time-consuming local philanthropic programs, many middle-class people assume that the welfare state will take care of the poor. This is especially attractive to the middle-class citizen who accepts the welfare-state myth that most of its cost derives from taxes on the rich. The facts are, of course, that the rich are adept at tax-exemption processes and the cost of the welfare state in many Western countries is paid for primarily by the taxes of the middle class and the working poor. The middle class also enjoys the privilege of having its own relatives provided for by the income-redistribution mechanisms of the welfare state, which means that many middle-class people no longer need concern themselves about their relatives in need. Their aged parents are thus cared for, even if this is accomplished by an impersonal check-disbursing mechanism. With the relegation of relatives to the welfare state, one need no longer concern oneself about them (including the need to visit them and to be involved with their affairs).

The welfare state also promotes the myth of efficiency, but the facts are that the bureaucracy directed by a distant centrality has rather a weak track record in terms of effective individuation. The social security system in the United States has experienced numbers of failures where the computer has declared beneficiaries dead, even when they were still alive and awaiting their support checks. This could hardly happen in a locally operated and directed program in close contact with its beneficiaries.

Still another reason for holding on to the welfare state is the fact that it covers over the cruel truth that an effective competitive economy requires that all its participants face the threat of destitution. That is the other side of the coin of free enterprise. Gilder (1981) stated that the most serious fraud of the welfare state creators is that they hide from adults and children in poverty a fundamental reality, namely, that to escape poverty and to live comfortably, those in poverty need most of all the incentive of threatening poverty. Palmer (1952) states that "one of the most disastrous of the many evil consequences of a welfare state is the prevalence of the crazy idea that the state owes a man a living whether he earns it or not" (p. 554). Hundreds of thousands of the young people in the welfare states have either not entered the labor market or withdrawn from it and, instead, have found ways to receive the benefits of the welfare state without being productive and without having to internalize the disciplines required of a self-sufficient worker. When their children (if they are not already the children of welfare parentage) reach adulthood, the welfare state and the governments sponsoring them can expect severe fiscal as well as social-control problems. Alan Wolfe (1983) says,

> [The] bureaucracy [of the welfare state] really is stifling. The assumption by the state of social responsibility means less of a sense of personal responsibility. Younger people, with little to look forward to, have created a culture of extreme alienation and even violence in Amsterdam, Copenhagen, and Stockholm. [There is] too little [personal] angst [and no personal responsibility]. [P. 22]

Before the days of the Industrial Revolution, people lived in small, close-knit communities where relationships were primary, personal, and affective. Inhabitants never went hungry as long as anyone had food, and no one was homeless as long as anyone had shelter. Public welfare was informal and was provided in the form of aid from one's nuclear family, one's extended family, and one's neighbors. This was the gemeinschaft community described by Tönnies (1963, chap. 1). No one in this community was ever lonely (but it is probable that many were frequently bored). Residents of the community held mutual obligations. All were more or less in a symbiotic relationship on economic, social, and political levels in that all were responsible to one another. The sick and the aged were taken care of by their families and the community. If anyone was in need, the community organized itself to provide help in finding work or in seeking out some other way to help them to help themselves. If any one person did not do his or her share of community work, he or she would soon hear about it from the neighbors.

When the marketplace community began to develop, a new form of community relationship took place. Tönnies called this the gesellschaft. In this type of community, the person's status no longer depended upon traditional membership in the community but upon salable skills, or available savings for entrepreneurship, or on skill in trading and organizing economic production. A person's value to the community was no longer based on past relationships with others and on community membership but, instead, was based alone on marketable qualities. The community emphasis was no longer on sentiment—it now rested entirely on economic values and on the business contract. Social control was no longer located in a circle of elders and kinship groups—it was now entirely vested in the interests of the marketplace. Thus, the focus of the gesellschaft was on profit (and loss) and on salable productivity. Anything that interfered with productivity and profitability was viewed as antagonistic to the free market and the functioning gesellschaft.

It was through the expansion of the free-enterprise system that productivity rose to such a level as to provide common people with goods (and life's necessities) of such quality and quantity as would have made kings and queens jealous in their time. This level of productivity was achieved primarily by the gesellschaft community patterns and the Protestant ethic, which were adopted by entire populations. When humane considerations arose (such as in the care of the sick, handicapped, and elderly), the policy adopted (as contained in the Elizabethan Poor Law) was to deal with them on a gemeinschaft basis. Thus, gemeinschaft relationships prevailed in the case of people unable to work, and gesellschaft relationships prevailed in all other cases. Under the Elizabethan Poor Law, aid to the poor was operated on a gemeinschaft basis with local responsibility, local discretion, local authority, and local funds, using the principle of lesser eligibility, the means test, and the responsibility of the client to rehabilitate and move toward self-sufficiency as soon as possible. When social-insurance plans were devised, these were provided on a preearned basis, using gesellschaft contractual relationships, and were made transferable among communities. Thus, the welfare system, which best functions on a noncontractual, individualized basis, was retained at a gemeinschaft level of relationships. The social-insurance system, which best functions on a quid pro quo, reciprocal exchange, mechanical, contractual basis, was placed at a gesellschaft level of relationships. Although other nations moved to place both welfare and social insurance on a gesellschaft basis, the Swiss elected to keep their services at the level of relationships where they were each best able to serve. The social-insurance relationship is primarily businesslike and is served best on a gesellschaft basis. The individual welfare client is best served at the local level, with local advice and help and within a realm of neighborly interaction.

In the gemeinschaft community, the family is supported by local sentiment and concern. Family policy is strengthened in local ordinances by direct concern for their well-being. Community concerns are protected in the cantonal legislature by local representatives to the cantonal parliament who are, in the last analysis, responsible to the voters of the community. In the United States, in comparison, family policy and community concerns have few elements of governmental protection.

The Swiss emphasis on gemeinschaft services at the local level and on gesellschaft services at the federal level of government is important. Many other Western nations have shifted almost all services to the gesellschaft format under a centralized bureaucracy. When this has happened, personal services in these nations (such as welfare, counseling, and sick and aged care) have proved faulty and unrealistic, perhaps because these are gemeinschaft services, which must necessarily be provided on a person-to-person basis unless they are to become what Rydenfelt (1981) has described as "hired love" (see Chapter 1). The care that a neighbor provides to a neighbor or a local social worker under local direction gives to a local client is far different from the mechanized services purchased or authorized by a central authority. Hired love is, in the last analysis, a contradiction in terms. A concern was expressed in Chapter 1 that the welfare state is replacing the family for major sectors of the population. The operation of a complex centralized bureaucracy assigned to perform a temporary gemeinschaft service is a contradiction in task assignment. It is equivalent to putting the fox in charge of the hen house. Without gesellschaft controls of supply and demand and profit and loss, a complex bureaucracy soon converts a temporary program into a permanent one and a personal service program into an impersonal one; and (to ensure its survival) it develops an incentive condition that makes of a limited clientele an expanding and dependent population sector. The centralization of gemeinschaft-type services by a federal authority is as faulty as the reverse, the decentralizing of gesellschaft-type services. Thus, it is suggested that the local and federal separation of services be reestablished for the better fulfillment of each, as has been proved in practice in Switzerland. The operations of gemeinschaft-type service on a gesellschaft basis opens up the possibility of clients who need personal attention being lost in the system. Clients not being answerable to their community because of their being served on a centralized, gesellschaft basis can become inured to continued benefits unless they are provided with the kind of relationship and supervision available only on a responsible local level. Just as child custody, child care, child neglect and abuse, adoptions, domestic relations, divorce, delinquency, and crime control are all matters of local concern and responsibility, so should welfare and the rehabilitation of welfare clients be located at the local gemeinschaft level.

The Swiss lesson is apparently that when local communities and local people are given their appropriate responsibilities and the authority to carry them out, they rise to the assignment. Because the central government is no great solver of problems, and because centralizing problems makes them only more difficult and complex and dependent upon volumes of administrative regulations needed to cover all national differences, the Swiss have learned to work on their problems where they exist—in the local community.

The question next to be answered is whether the Swiss experience is unusual. Are the Swiss solutions inapplicable to other nations? This requires examination.

Is Switzerland a special case, a fortunate accident, a small exception? When one considers that Switzerland is as large or larger than many other European nations, including many welfare states (such as the Netherlands, Belgium, and Norway), one may conclude that Switzerland is comparable in size. Is the Swiss culture so imbued with the religiously oriented Protestant ethic as to repress welfare dependency? If so, one should consider other countries with similar or even stronger Calvinist influences that are welfare states, among them the Netherlands, Norway, and Sweden.

Another explanation for Switzerland as a special exemption might be founded on its historical development. This argument would hold that Switzerland has not yet caught up with the ills of the rest of the industrial world and that in time the Swiss can be counted on to drop their inhibitions against dependency and pride in self-reliance. This argument also can be countered with the facts that, aside from a short-lived restrained youth protest and some limited evidence of premarital coupling, there is little serious social disorganization in the Swiss society. The social institutions of Switzerland are still intact. There are almost no slums. The social structure is stable. The family is still important. Open adultery is unknown rather than condoned, as it is in many other countries. The Swiss work hard, and their children have lives directed toward goals of self-sufficiency and social responsibility. From these indicators, it would be difficult to portend a breakdown of the Swiss society such as might well have been predicted in many other nations.

Another argument against using Switzerland as a comparative model might point to the strong Swiss culture, which presumably serves to prevent welfarization and related social ills. The Netherlands and the Scandinavian nations also had cultures that strongly supported the Protestant ethic, but in these nations welfare-state disincentives have seemingly dissolved the ethical imperative of self-sufficiency. As was noted in Chapter 1, the welfare state can and does replace the family, community institutions, and the cultural imperatives and restraints. Only in Switzerland has this not happened—but Switzerland does not have a welfare state. Therefore, if welfare works in Switzerland, it is not because of the Swiss culture but because of the Swiss welfare system and the manner in which Swiss people deal with poverty.

A final argument against validating the Swiss example is the claim of homogeneity. This claim is that Switzerland is made up of a homogeneous population, unlike many other nations with welfarization problems. But, is Switzerland really homogeneous? With four different languages, two major religions and many schisms appearing within them, a multiplicity of political parties, and opinions matching those in many other nations, is Switzerland a homogeneous nation? The answer, after analysis, is in the negative.

Seligman's (1975) work on learned helplessness can be used to explain the effect of welfare-state security provisions on those in the population who are not yet prepared to live the life of risk and self-testing that is the basis of the free-enterprise system. A person will draw back when faced with the risk of failure (which real opportunity provides) and will be attracted and psychologically rewarded by readily accessible and sizable programs of income redistribution. In a sense, as the population accepting the programs of income redistribution grows, this creates a culture of its own, making not learning, not training, and not working an acceptable way of life. By not providing rewards for learned helplessness and by enforcing social control against such behavior, Switzerland provides an antiwelfarization example that is replicable in other countries.

And so we are forced to accept a last set of explanations; namely, that Switzerland presents the world with a carefully designed set of welfare policies, which is reinforced by effective patterns of parenting, schooling, vocational preparation, employment, and local governmental structures and supported by an involved and concerned population. These add up to effective prevention and control of welfare dependency and a host of associated ills.

That the Swiss have low rates of welfare dependency, delinquency, and crime is well known. They have demonstrated that these and many other urban ills are in truth everybody's business. The reason that the non-Swiss Western city has become unsafe and unlivable is that so many of us rely on someone else to take care of social problems. But the Swiss have found that only in the local community can everybody be brought to concern themselves with the problems they must face. They have demonstrated that social problems can be resolved only with the active support of the entire community and that only in the local community can social control ensure everyone's involvement. In a sense, people get the kind of community life they work for: if they behave as if they are not concerned, they are soon surrounded with urban ills and dangers; if everyone is actively involved and responsible, then all can achieve a safe and livable community.

ARE THERE LESSONS IN THE SWISS EXAMPLE?

Does the Swiss example provide lessons for the troubled welfare states and semi-welfare-state nations? Can the Swiss experience help solve the problems of welfarization? How has Switzerland avoided the welfarization problem?

We can choose to focus on welfare dependency as one of the problems, but careful analysis leads one to the admission that welfarization is only one of a complex of interrelated social dysfunctions that Switzerland has apparently avoided. It is as if all the Swiss had gotten together and agreed to do everything possible to make their members productive. Whether this complex of social problems has been avoided because of careful Swiss social planning or whether these problems have been avoided because of a complex of Swiss norms and values is not known. In any case, the avoidance of welfare-state methodology made it possible for Switzerland to provide conditions for its population that surpass those in many of the nations that claim to be welfare states.

What are the norms, values, and methods by which Switzerland achieved this situation? It is interesting to note that Switzerland's first constitution was modeled after the first U.S. constitution. Thus, Switzerland is a confederation of states, just as was the United States. Although the United States shifted to a strong federal government, Switzerland retained the confederation model. As a result of the shift, most U.S. power has moved to Washington, D.C., and urban problems are generally at the mercy of federal support.

As a result of the Swiss model of government, the Swiss respondents to the author's study reported that the best-informed people are usually elected to local offices and that usually the least corrupt and most concerned in dealing with the unfortunate become the local authorities. The pathway to local leadership is comparatively open in the Swiss community. Because everyone is watching everyone on the local scene and because financial accounts are legally open, there is very little fraud. In an interview with Hans Peter Tschudi, former president of Switzerland, the author learned that he had gotten his start in Swiss politics by working as a commissioner in the Basel community on their buildings and works program. In over twenty years of service on this commission, not one instance of fraud was reported.

Apparently, experience has shown that only at the local level is it possible for everyone to know one another. It is also possible at this level of ask questions and to receive clear answers. It is possible at the local level to understand the issues and the people affected. And at the local level the citizen either gets involved (in the issues) or has to suffer with the results of someone else'e activity.

Obviously, it would not be possible for the other Western nations to completely decentralize the resolution of social problems. But probably the first lesson from the Swiss experience would be to relegate as many problems and tax bases as possible back to smaller local units of government where the citizenry can get involved. The apathetic or dormant local U. S. community would soon awake when it finds that it holds both the responsibility and the assets related to the issues.

Another basic difference between Switzerland and other Western nations relates to the power of the federal legislature and the federal courts. On the Swiss scene, the federal courts have never entered into the question of the validity or invalidity of legislative or executive actions. The Swiss federal courts serve only as a final arbiter in individual conflicts but do not enter into the creation of law by class actions. That, the Swiss believe, is the job of the legislatures, and in the last analysis it is up to the voters either by referendum or initiative. Thus, a community or canton may act in accordance with the will of its voters, and as long as it does not interfere with other communes or cantons or the cantonal constitution, it may continue this policy despite the fact that its actions do not parallel those of other communities and cantons. Obviously, the courts cannot be redesigned in other nations, but as local autonomy based on local financing takes hold, it may become possible for local programs to function with less involvement of federal courts. Swiss legislation always has a central test question asked before anything is submitted to the voters for approval: Does this law grant rights for something without holding the beneficiary responsible for his or her actions? Does this law hold someone responsible for some act or behavior without matching that responsibility with its related rights? The Swiss, by long experience, have learned that to grant rights without responsibility, or to place responsibility without commensurate rights, will only end with unanticipated, distorted, and often counterproductive consequences in the behavior of the affected persons. For each of the rights or benefits granted, there is an established quid pro quo to be performed. This could easily be introduced into other nations by autonomous local welfare programs using their own tax base and without federal involvement. It should be noted that Swiss elected officials at all levels of government are closely attuned to the complex of Swiss ethics involving honesty, self-reliance, hard work, the norm of reciprocity, local control, and family responsibility. No law or regulation is passed that may in any way operate counter to these ethics, except in the instance of the aged or sick, who cannot fend for themselves. Even in such instances, the government seeks to use private programs of aid if possible, and great care is exerted to prevent federal or cantonal intrusions into local affairs. The problem of concubinage (persons living together without marriage) was seriously considered by all levels of government because it poses a possible threat to family life and adequate child socialization. But after much discussion, it was left to the cantons to act, and in seven cantons the practice is still illegal.

The Swiss have learned that culture can operate counter to the law, in which case the law becomes vacuous or the ethic is weakened. Culture can also operate to support the law, in which case the law is strengthened and the society remains stable. This is the fortunate condition. Had the Swiss society passed laws that operated counter to the prevalent ethic, it is quite possible that counterproductive social change might well have occurred and Switzerland would today be faced with many of the same problems as other Western nations.

The Swiss pattern of strict separation between the social insurances and welfare services is well founded. The pattern of providing for local welfare services on a rehabilitative, personalized basis has proved effective.

The lessons of Switzerland for other Western nations are clear. Social problems cannot be resolved without provision of appropriate employment, without control of immigration, without well-planned and earned social insurances, without promotion of family responsibility, without appropriate child socialization, without effective schooling aimed at academic excellence, without vocational preparation, and without cooperative support of such efforts by the domestic relations courts.

Only in the local community or neighborhood is it possible for agencies of social service and social control to keep close touch with one another in the individualized treatment of clients. Even then, the agencies must necessarily be autonomous and not constrained by national regulations, which preclude the design and administration of unified case planning. Switzerland has demonstrated that such coordination at the local level can be achieved and effectively carried out. Swiss welfare is local, cooperative, and nonpoliticized. Finally, as the Swiss have amply demonstrated, the control and prevention of welfarization and its related social ills requires involvement of the citizenry in the local community.

The resolution of welfarization problems and related urban ills of the Western nations cannot be quickly achieved. It may take a nation as long to solve the problems as it took to create them.

Etzioni (1983), in his "immodest agenda" for the United States, concludes that the local community is the only viable force capable of holding the society together. The local community would be able to do this because only the local community can direct the shared concerns of the constituent members. Etzioni laments the rise of centralized government and the "me" generation culture, which have furthered the weakening of local dynamics. Etzioni's emphasis on the need for mutuality, commitment to others, shared concerns for civility, and sensitivity to preserve basic values and institutions is matched only on the local Swiss scene.

Fuchs (1983) points to the destructive effect the growth of unwed motherhood and the proliferation of one-person households in the United States has on the family and on children's socialization (p. 221). Mitscherlich (1970) indicates (pp. 141-42) that with the absence of the father, the teenager who has never known his or her father develops an ego that listens only to basic urges. "His (or her) whole perception of the environment is in the ruthless service of his (or her) instinctual values," having never developed the socializing identification based on the necessary dependable figures in the environment. The resultant fantasy models never call for moderation or self-control. In the absence of direct and immediate preparation for instruction drawn from competent and complete parenting under the paternal eye, the youthful contemporaries orient themselves by each other and their pre- or anti-social patterns of behavior (p. 149). Thus, the educational rearing of children (such as it is) and the substantive financial support of the childlike mothers and their children become the responsibility of the schools and public welfare agencies, creating, in turn, the huge bureaucratic apparatus of the paternalist state and concealing an enormous amount of waste built around a system that operates on the principle of avoiding responsibility and evading knowledge of and control of its effects (p. 227). Hirschi's studies of crime and the family indicate that the percentage of households headed by women "is a major and powerful predictive factor of crime rates" (p. 61). The children from intact families have lower rates of crime than children from broken or reconstituted homes (Hirschi 1983, p. 62).

On the basis of these indicators alone, a social policy that either weakens the institutions of the intact family and the effective community or serves to provide weak incompetent substitutes for them would have to be seriously questioned. A welfare system that makes dropping out of school easier, that makes nonpreparation for employment and career routine, that makes irresponsible fathering of illegitimate children normative, that makes unmarried motherhood an accepted rite of passage, and that turns hundreds of thousands of children onto the ghetto and suburban streets for rearing by their premoral peers surely requires revision. It would be hoped that the Swiss way of welfare, which avoids these ills, would be examined as a basis of comparison and as a possible model for revision.

Moynihan (1982) issued a call to Americans to devise social institutions that deal with the vulnerability of the nuclear family in a postindustrial economy. He said that "we must provide ways to support dependent children without introducing incentives to child abandonment. We need to find institutions that generate norms and encourage self-reliance among adults. Most of all . . . we need to recognize that the problem is growing" (p. 20). Obviously, Switzerland has demonstrated that the problem is resolvable.

Twice before in recent history, the United States has faced the difficulties of solving a problem without political manipulations, simplistic argumentation, and resort to ideologies. Only twice recently has the United States sought to attain a solution without concern over structural barriers and constitutional complications. On these occasions, the country acted as one entity—in recovering from the Great Depression and in winning the war against Hitler. One was an economic crisis, the other a military crisis. The United States now faces a social crisis of the family, the community, and the socialization of its children. An entire population is at risk.

The United States has the technology, the resources, and the social-psychological knowledge to eradicate poverty and to remove the blight and sickness of its cities. Whether or not this will be attempted and finally achieved only history will reveal.

REFERENCES

Anderson, Digny, June Lait, and David Marsland. 1981. Breaking the Spell of the Welfare State. London: Social Affairs Unit.

Etzioni, Amatai. 1983. An Immodest Agenda: Rebuilding America before the Twenty-first Century. New York: McGraw-Hill.

Fuchs, Victor R. 1983. How We Live. Cambridge, Mass.: Harvard University Press, p. 221.

Gilder, George. 1981. Wealth and Poverty. New York: Basic Books.

Hirschi, Travis. 1983. "Crime and the Family." In Crime and Public Policy, edited by James Q. Wilson, pp. 53-68. San Francisco: Institute for Contemporary Studies.

Mitscherich, Alexander. 1970. Society without the Father. New York: Schocken Books.

Moynihan, Daniel P. 1982. "One-third of a Nation: How Will America Care for Its Children?" New Republic, June 9, pp. 18-21.

Palmer, Cecil. 1952. The British Socialist Ill-Fare State. Caldwell, Idaho: Caxton.

Rydenfelt, Sven. 1981. The Rise and Decline of the Swedish Welfare State. Lund: Nationalekonomiska Institutionen, Lunds Universität.

Seligman, M. E. P. 1975. Helplessness: On Depression, Development, and Death. San Francisco: W. H. Freeman.

Tönnies, Ferdinand. 1963. Community and Society: Gemeinschaft and Gesellschaft, edited by C. P. Loomis. New York: Harper.

Wolfe, Alan. 1985. "The Death of Social Democracy." New Republic, February 25, pp. 21-23.

INDEX

Aarau welfare (see also Swiss
local welfare systems): and absent fathers, 146; and aid to clients, 143-44; and appeals of clients, 143; caseload of, 142, 145;
cases of, difficult, 145-46; and
divorced fathers, 147; drawbacks
of, 146-47; and drug-counseling
center, 117; and employment,
143; and grants to clients, 143;
and homemakers, 147; and intracantonal contract, 142; policies
of, 142-44; procedures of, 144;
responsibilities of, 144; volunteers in, 147; and youth advice
agency, 144-45
accident insurance, 82
AFDC: clients, 16; expectation
about, 3; growth of, 9-10; and
natural families, 2-3; and Swiss
family allowances, 83
AHV (see old-age and survivors'
insurance)
Aid to Family with Dependent Children (see AFDC)
alcoholism, 175-76
Aldous, Joan, 2
Allgemeine Sozialhilfe (see Basel
welfare)
Anderson, C. LeRoy, 7
Anderson, Digny, 186-87
ATD Vierte Welt, 173-74
Auletta, Ken, 174

Banfield, Edward C., 2-3
Basel welfare (see also Swiss
welfare systems): and aid for clients, 141; and appeals of clients,
141-42; caseload of, 140; and
drug-addiction cases of, 142; policy of, 141; problems of, 140

Battle, Ester S., 7-8
Beeghley, Leonard, 3
Belgian welfare, 36-38
Benoit, Jean, 33
Berelson, Bernard, 181
Bern welfare (see also Swiss local
welfare systems): and appeals
of clients, 135; comparison of,
to Zurich welfare, 130-31; employment goals of, 134-35; and
follow-up of clients, 134; and
grants to clients, 135; and housing for aged, 134; mandate for,
131; personnel of, 134; policies
of, 131-32; purpose of, 131; report of, 1980, 134; services of,
132-34; and unmarried couples,
135
Beveridge model, 72
Bismarck model, 72
British welfare, 20-22
Büchi, O. A., 115, 117

case poverty, 18
Castaing, Michel, 33
Castellino, Onorato, 34-35
Charles, Jean-François, 75
Chilman, Catherine S., 2-3
civil defense, 166
Clark, Gardner, 55, 57
class poverty, 18
Clinard, Marshall B., 57, 164,
176, 178
Coleman, John R., 24
community, 160
community activities, 166
compulsory occupational retire-
ment scheme, 74-76
Conference of Cantonal Welfare
Directors, 117
Cook, Dan, 32

Danish welfare, 35–36
definition of situation, 103
divorce courts, 167–68
domestic relations courts: divorce, 167–68; operation of, 168; and Swiss life-styles, 169–70; and Western life-styles, 168–69
Dornbusch, Sanford M., 8
Drouin, Pierre, 33
drug abuse, 178–79
Dunn, Nell, 22
Dutch welfare, 22–25

education: ideology of, 161; parent involvement in, 163; and reciprocity norm, 162; rewards of, 163–64; and socialization, 161–62; teachers in, 162–63
Ellwood, David T., 2, 16
employment, 164–65
Ergänzungsleistungen provision, 78–79, 120
Ernst, Urs, 89
Etzioni, Amatai, 195
European welfare: administrators of, 109, 112; in Belgium, 36–38; client-worker relationships in, 118; and definitions of poverty, 17–18; in Denmark, 35–36; in France, 30–33; in the FRG, 25–30; in Ireland, 36–38; in Italy, 33–35; and national government, 107, in the Netherlands, 22–25; past view of, 2; principles of, 129; social-control functions of, 118; and social-insurance policies, 18; in Sweden, 19–20; in United Kingdom, 20–22

family: allowances for, financial, 83–84, 89–90, 158–59; canton support of, 156–57; child rearing in, 153; cohesion of, 152–53; community support of, 154–56; divorces in, 150, 152; and education of

children, 154, 156–58; future of, 155–57; housing of, 155; illegitimate children in, 150–51; Lüscher report on, 1980, 151; marriages in, 150, 152; policy of, 171–72; responsibilities of, 153–54; unmarried mothers in, 151
family allowances, 83–84, 89–90, 158–59
Federal Act of 1911, 80, 82
Federal Act of March 1981, 82
Federal Social Insurance Act, 25
Ford, Thomas R., 7
Ford Foundation, 13
Forman, Rachel Zinober, 21, 174
French welfare, 30–33
Frenkel, Max, 60–61
Fuchs, Victor R., 195
Future That Doesn't Work: Social Democracy's Failure in Britain, 21

Galbraith, John Kenneth, 52
Gastarbeiters, 99
gemeinschaft community, 188–90
Geneva welfare (see also Swiss local welfare systems): for aged, 136; establishment of, 135–36; and grants to clients, 137; for handicapped, 136; and loans for clients, 137–38; and out-of-canton clients, 138–39; personnel in, 139–40; for refugees, 138; and relatives' responsibility, 136; report of, 1982, 137; for single parents, 136–37; and unmarried couples, 139; and worker placement, 139
German welfare, 25–30
gesellschaft community, 189
Gilder, George F., 3–4, 188
Goodman, Paul, 18
Goudswaard, K. P.: and British disability programs, 22; and

Danish expenditures, 35; and French disability programs, 31; and German disability programs, 28, 30; and Italian GDP, 34
Grenzegängers, 95
Gross, Peter, 161
Growing Up Absurd, 18
Gruner, Erich, 162
Gruntwig, N. F. S., 36
guilds, social, 166

Halberstadt, Victor: and British disability programs, 22; and Danish expenditures, 35; and Dutch welfare state, 23-24; and French disability programs, 31; and German disability programs, 28, 30; and Italian GDP, 34
Handbook of Social Policy, 114
Hauser, Jürg A., 123
Haveman, R. H., 22, 28, 30-31
Heckscher, Gunnar, 20
Herzog, Elizabeth, 7-8
Hill, Reuben, 2
Hirsch, Joachim, 36
Hirschi, Travis, 196
Hoffmann-Novotny, Hans-Joachim, 89, 150, 175
Homans, George, 3-4
Horlick, Max, 75, 78
Horney, Karen, 53
Hospice Général (see Geneva welfare)
Huntford, Roland, 20

immigrants (see also Swiss immigration policy): in France, 32-33; in the FRG, 26-27; in Geneva, 138; in Switzerland, 94-95; in the United States, 10-11
individual poverty, 18
Institutes for Family and Marriage, 170
invalidity insurance: background of, 76; benefits of, 76; costs of,

76; rehabilitation services of, 76; study of, 77-78; supplementary benefits of, 78-79; and U.S. system, 76-77
Irish welfare, 36-38
Italian welfare, 33-35
IV (see invalidity insurance)
Izbieki, John, 21

Jaffee, Stephen, 176-77
Janssen, Martin C., 74, 79
Jenkins, Peter, 21
Juvenile Control Office, 159

Keller, Theo, 112
Kluckhorn, Florence, 7
Kramer, Jane, 21-22, 104

Lait, June, 186-87
LeBlanc, Bart, 34-35
Liebfried, Stephan, 26
Luck, J. Murray, 58, 61-62, 152
Lunn, John H., 21
Lüscher, Kurt K., 151, 161

Madge, Nicola, 4
marginal poor, 8, 11-12
Marshall, Tyler, 36
Marsland, David, 186-87
mental health services, 170-71
military service, 165-66
Miller, Walter, 7
Ministry of Social Affairs and Public Health, 170
Mitscherlich, Alexander, 196
Moynihan, Daniel P., 2, 196
Müller, Heinz H., 74, 79
Müller, Richard, 176
Murray, Charles, 17

National Assistance Act, 1965, 23
Neue Zürcher Zeitung, 89
neurotic exemption, 53
New York study, of U.S. welfare, 4-7

Oberhasli district welfare program, 117

old-age and survivors' program: benefits of, 72-73; and Beveridge model, 72; and Bismarck model, 72; funding of, 73; insurees of, 73-74; problems of, 74; purpose of, 71-72; supplementary benefits of, 78-79

Painton, Frederick, 28, 36
Palmer, Cecil, 188
Parry, Wilfred, 21
persistent poverty, 17
Poor Cow, 22
poverty: class, 18; European definition of, 17; in France, 33; in the FRG, 26, 28; individual, 18; interactional definition of, 17-18; intergenerational, 1-4; persistent, 17; relative deprivation, definition of, 17; Strang's models of, 17-18; structural, 18; in Switzerland, 55; U.S. models of, 8; in Zurich, 123
poverty trap, 4-5
Pro Familia, 171
professional pensions system (see compulsory occupational retirement scheme)
Pro-Infirmis, 113-14, 120
Pro-Juventute, 113-14
Pro-Senectute, 113-14, 120
Pruitt, Walter, 7
psychiatric facilities, 170-71
public assistance (see Swiss local welfare systems)

Ramuz, C. F., 56
Rattner, Julian B., 7-8
Reidi, Victor, 178
Reisman, David, 162
Reist, W., 175-76, 178
residual poor, 8, 12
Ritter, Verena, 161

Rosa, Jean-Jacques, 32
Rutter, Michael, 4
Rydenfelt, Sven, 19-20, 190

Salamon, Lester M., 2
Schanche, Don, 34
Schlamp, Frederic T., 7
Schmid, Carol L.: and cohesiveness of Swiss families, 152; and foreign workers in Switzerland, 100, 103; and immigration patterns in Europe, 97; and politics, Swiss view of, 60
Schmidt, Joseph, 28
Schneiderman, Leonard, 7
Schweizer, Willy: and average annual income, Swiss, 89; and GNP, Swiss, 57; and military service, Swiss, 165; and retirement payments, Swiss, 84; and sickness insurance, Swiss, 80; and social insurance system, Swiss, 66-67; and unemployment payments, Swiss, 83, 165
Seewandono, Iwan, 17
self-help organizations, 116
Seligman, M. E. P., 192
service clubs, 166
Sharff, Jagna Wojcicka, 7
Sheehan, Susan, 4-8, 174
sickness insurance, 79-82
Siegenthaler, Hansjorg, 58
Simanis, Joseph G., 24
Skolnik, Alfred M., 75, 78
Smalley, Ruth, 6
Smarz, Zofia, 28
social security (see European welfare; Swiss welfare; U.S. welfare)
Social Service Office, 159
Sozialamt, 159
SSI, 16, 77
Stack, Carol B., 3
Steiner, Gary A., 181
Stevenson, Harold W., 163

Stone, Robert C., 7
Stovall, Sten, 36
Strang, Heinz: and causes of poverty, 112; and models of poverty, 17–18; and welfare in the FRG, 28–30
structural poverty, 18
Summers, Lawrence H., 2, 16, 77
SUVA, 82
Swedish welfare, 19–20
Swiss Conference of Public Welfare, 117
Swiss controls of welfarization (see also Swiss factors in welfarization): community, 160; and dependency caseloads, 172–75; domestic relations courts, 167–70; employment, 164–65; family, 150–59; family policy, 171–72; military service, 165–66; psychiatric facilities, 170–71; schooling, 161–64; youth authority, 164
Swiss factors in welfarization (see also Swiss controls of welfarization): alcoholism, 175–76; drug abuse, 178–79; youth riots, 179–82
Swiss Federal Advisory Commission for Youth Affairs, 179
Swiss immigration policy: comparisons of, with other policies, 93–94; criticism of, 94–95; debate of, in 1970s, 103; and education of foreign children, 102–3; and employment procedures, 94; and immigration patterns of other countries, 97–99; and influx of foreign workers, 95–98; and integration of foreign workers, 103–4; and position of foreign workers, 102; and poverty prevention, 54; and recession, employment, 96–97; and stability

of foreign workers, 100–1; success of, 104–5
Swiss local welfare systems (see also Aarau welfare; Basel welfare; Bern welfare; Geneva welfare; Zurich welfare): and administrator's view of poverty, 111–13; and aid to clients, 109, 147; client-worker relationships in, 117–20; comparisons of, to other systems, 107, 109, 147–48; confidentiality in, 110–11; coordination of, 116–17; expenses of, 120, 123; funding for, 115–16; and gemeinschaft programs, 120; and local government, 107–8, 147; policies of, common, 108–10; and self-help groups, 116; and voluntary agencies, 113–15
Swiss National Accident Insurance Institute, 82
Swiss social insurance (see also Swiss welfare): accident insurance, 82; belief system underlying, 66, 70; comparison of, to other systems, 51; compulsory, 66; compulsory occupational retirement scheme, 74–76; design of programs of, 70–71; family allowances, 83–84, 89–90; invalidity insurance, 76–78; old-age and survivors' insurance, 71–74; operation of, 66; policy of, 53–54; and principles of welfare state, 51; sickness insurance, 79–82; supplementary benefits of programs, 78–79; unemployment compensation, 82–83; voluntary 66
Swiss welfare (see also Swiss controls of welfarization; Swiss local welfare systems; Swiss social insurance): cases of dependency on, 172–75; and communities, 160; comparison of, to

other welfare systems, 49–51; costs of, 120; and definition of welfare state, 48–49, 51; factors of dependency on, 172–75; gemeinschaft services of, 190; and government, 55–56; images of, 48; and immigrants, 54–55; lessons of, 190–92, 195; plan for, 51; policies of, 53–54; and poor, 52–53; and relatives' responsibility, 109–10; and social policy, 59–62; success of, 52, 193–97

Swiss welfarization (see Swiss controls of welfarization; Swiss factors in welfarization)

Switzerland (see also Swiss welfare): demography of, 55; economy of, 56–58; geography of, 55; government of, 55–56; immunity of, to social ills, 58–59; income distributions in, 84, 89–90; social policy of, 59–62

Taki, Theodoracopulos, 179
Thomas, W. I., 103
Tifflin, John, 23–24
Tönnies, Ferdinand, 188–89
Townsend, Peter, 21
transitional poor, 8, 10–11
transmitted deprivation, 17, 21
Treffpunkt, 26–27
Trimborn, Harry, 176–77
Tschudi, Hans Peter, 82–83
Tuggener, Hans, 152
tutto per tutti, 33
Tyrell, R. Emmett, Jr., 21

Underclass, The, 174
unemployment: in Belgium, 36; in Ireland, 36; in Italy, 34; in Switzerland, 50, 50–51; in the United Kingdom, 21
unemployment compensation, 82–83
U.S. welfare: administrators of, 109, 112; anthropological study of, in East Harlem, 7; client-worker relationships in, 118; controversy over, 12–13, 16; and delinquency, 4, 7–9; dependency on, 1–4, 16; and education of children, 5–6, 13, 16; effects of, 16–17; expectations about, 3–4; growth of, 9–10; income redistribution view of, 12; and invalidity insurance, 76–77; issues in, 13; and lessons from Swiss welfare, 195–97; and national government, 107; past view of, 2; and poverty models, 8, 10–12; and poverty trap, 4–5; principles of, 127, 129; problems of, 1–2, 13–14; and retirement insurance, 75–76; Sheehan study of, in New York City, 4–7; social-control functions of, 118; stigma of, 5; structural functional view of, 2; and worker-client relationships, 6–7

upward mobility, 181
Van de Castle, John, 7
Van Doorn, Jacques, 17, 24
Vaud Canton welfare, 117
voluntary agencies, 113–15

Wagner, R., 175–76, 178
Walliman, Isidor, 31–32, 104
welfare (see European welfare; Swiss welfare; U.S. welfare)
welfare dependency (see also Swiss controls of welfarization; Swiss factors in welfarization): in Denmark, 35–36; in France, 32–33; in the FRG, 26–27; in Italy, 34; in the Netherlands, 23–24; problems of, 1–2; in Sweden, 21; in the United Kingdom, 22; in the United States, 2–4
Welfare Mother, A, 4–7, 174

welfare state (see also European welfare, Swiss welfare; U. S. welfare): perceptions about, 186-88; rise of, 188-90; success of, in Switzerland, 191-92

welfarization (see Swiss controls of welfarization; Swiss factors in welfarization)

West, Donald J., 22

Wilensky, Harold L., 48-49, 51

Wilson, William Julium, 2

Wolfe, Alan, 188

Wolfe, B. L., 22, 28, 30-31, 34

Work Placement Bureau of Geneva, 139

Wright, Catherine, 21

youth advice agency, 144-45

Youth Authority, 155, 159, 164

youth riots, 179-82

Yverdon welfare, 117

Zurich welfare (see also Swiss local welfare systems): comparison of, to other cities' welfare, 130; listing of services of, 114; policies of, 125-26; population on, 123-24; principles of, 126-27, 129; reports on services of, 129-30; role of, 130

ABOUT THE AUTHOR

RALPH SEGALMAN is a Professor of Sociology, California State University-Northridge, where he has taught since 1970. He previously was on the faculty of the University of Texas at Austin and El Paso. Prior to his academic career, he served for 25 years in various capacities as a professional social worker. He is the author of Dynamics of Social Behavior and Development (1978), The Deviant, Society, and the Law (1978), and (with Asoke Basu) Poverty in America: The Welfare Dilemma (1981). He has also authored some 49 published articles in various journals.

Professor Segalman received his bachelor's and Master of Social Work degrees from the University of Michigan and his Doctorate in Social Psychology from New York University. He is an Honor's Day Scholar of New York University, a fellow of the American Association for the Advancement of Science, and a recipient of the CSUN-Scholars award and, recently, of a meritorious performance faculty award from California State University-Northridge.